Quantum Computing by Practice

Python Programming in the Cloud with Qiskit and IBM-Q

Second Edition

Vladimir Silva

Apress®

Quantum Computing by Practice: Python Programming in the Cloud with Qiskit and IBM-Q, Second Edition

Vladimir Silva
CARY, NC, USA

ISBN-13 (pbk): 978-1-4842-9990-6
https://doi.org/10.1007/978-1-4842-9991-3

ISBN-13 (electronic): 978-1-4842-9991-3

Managing Director, Apress Media LLC: Welmoed Spahr
Acquisitions Editor: Melissa Duffy
Development Editor: James Markham
Coordinating Editor: Gryffin Winkler

Cover designed by eStudioCalamar

Cover image by WangXiNa on Freepik (https://www.freepik.com/)

Distributed to the book trade worldwide by Apress Media, LLC, 1 New York Plaza, New York, NY 10004, U.S.A. Phone 1-800-SPRINGER, fax (201) 348-4505, e-mail orders-ny@springer-sbm.com, or visit www.springeronline.com. Apress Media, LLC is a California LLC and the sole member (owner) is Springer Science + Business Media Finance Inc (SSBM Finance Inc). SSBM Finance Inc is a **Delaware** corporation.

For information on translations, please e-mail booktranslations@springernature.com; for reprint, paperback, or audio rights, please e-mail bookpermissions@springernature.com.

Apress titles may be purchased in bulk for academic, corporate, or promotional use. eBook versions and licenses are also available for most titles. For more information, reference our Print and eBook Bulk Sales web page at http://www.apress.com/bulk-sales.

Any source code or other supplementary material referenced by the author in this book is available to readers on GitHub (https://github.com/Apress). For more detailed information, please visit https://www.apress.com/gp/services/source-code.

Paper in this product is recyclable

To my dear parents, Manuel and Anissia, and beloved siblings, Natasha, Alfredo, Sonia, and Ivan.

Table of Contents

About the Author

Vladimir Silva was born in Quito, Ecuador. He received a System's Analyst degree from the Polytechnic Institute of the Army in 1994. In the same year, he came to the United States as an exchange student pursuing an M.S. degree in Computer Science at Middle Tennessee State University. After graduation, he joined IBM as a software engineer. His interests include Quantum Computing, Neural Nets, and Artificial Intelligence. He also holds numerous IT certifications including OCP, MCSD, and MCP. He has written many technical books in the fields of distributed computing and security. His previous books include *Grid Computing for Developers* (Charles River Media), *Practical Eclipse Rich Client Platform Projects* (Apress), *Pro Android Games* (Apress), and *Advanced Android 4 Games* (Apress).

About the Technical Reviewer

 **Jason Whitehorn** is an experienced entrepreneur and software developer and has helped many companies automate and enhance their business solutions through data synchronization, SaaS architecture, and machine learning. Jason obtained his Bachelor of Science in Computer Science from Arkansas State University, but he traces his passion for development back many years before then, having first taught himself to program BASIC on his family's computer while still in middle school.

When he's not mentoring and helping his team at work, writing, or pursuing one of his many side-projects, Jason enjoys spending time with his wife and four children and living in the Tulsa, Oklahoma, region. More information about Jason can be found on his website: https://jason.whitehorn.us.

Introduction

The Quantum Computing Revolution

I wrote this book to be the ultimate guide for programming a quantum computer in the cloud. IBM has made their quantum rig (known as the IBM Quantum) available not only for research but for individuals, in general, interested in this exciting new field of computing.

Quantum computing is gaining traction and now is the time to learn to program these machines. In years to come, the first commercial quantum computers should be available, and they promise significant computational speedups compared to classical computers. Nowhere is this more apparent than in the field of cryptography where the quantum integer factorization algorithm can outperform the best classical solution by orders of magnitude, so much so that a practical implementation of this algorithm will render current asymmetric encryption useless.

All in all, this book is a journey of understanding. You may find some of the concepts explained throughout the chapters difficult to grasp; however, you are not alone. The great physicist Richard Feynman once said: "*If somebody tells you he understands quantum mechanics, it means he doesn't understand quantum mechanics.*" Even the titans of this bizarre theory have struggled to comprehend what it all means.

I have tried to explore quantum computation to the best of my abilities by using real-world algorithms, circuits, code, and graphical results. Some of the algorithms included in this manuscript defy logic and seem more like voodoo magic than a computational description of a physical system. This is the main reason I decided to tackle this subject. Even though I find the mind-bending principles of quantum mechanics bizarre, I've always been fascinated by them. Thus, when IBM came up with its one-of-a-kind quantum computing platform for the cloud and opened it up for the rest of us, I jumped to the opportunity of learning and creating this manuscript.

Ultimately, this is my take on the subject, and I hope you find as much enjoyment in reading it as I did writing it. My humble advice: Learn to program quantum computers; soon they will be ever present in the data center, doing everything from search and simulations to medicine and artificial intelligence. Here is an overview of the manuscript's contents.

Chapter 1: Quantum Fields: The Building Blocks of Reality

It all began in the 1930s with Max Planck's reluctant genius. He came up with a new interpretation for the energy distribution of the light spectrum. He started it all by unwillingly postulating that the energy of the photon was not described by a continuous function, as believed by classical physicists, but by tiny chunks, which he called *quanta*. He was about to start the greatest revolution in science in this century: *quantum mechanics*. This chapter is an appetizer to the main course and explores the clash of two titans of physics: Albert Einstein and Niels Bohr. Quantum mechanics was a revolutionary theory in the 1930s, and most of the scientific establishment was reluctant to accept it, including the colossus of the century: Albert Einstein. Fresh from winning the Nobel Prize, Einstein never accepted the probabilistic nature of quantum mechanics. This caused a rift with its biggest champion: Niels Bohr. The two greats debated it out for decades and never resolved their differences. Ultimately, quantum mechanics has withstood 70 years of theoretical and experimental challenges, to emerge always triumphant. Read this chapter and explore the theory, experiments, and results, all under the cover of the incredible story of these two extraordinary individuals.

Chapter 2: Richard Feynman, Demigod of Physics, Father of the Quantum Computer

In the 1980s, the great physicist Richard Feynman proposed a quantum computer. That is a computer that can take advantage of the principles of quantum mechanics to solve problems faster. The race is on to construct such a machine. This chapter explores, in general terms, the basic architecture of a quantum computer: qubits – the basic blocks of quantum computation. They may not seem like much but they have almost magical properties: Superposition, believe it or not, a qubit can be in two states at the same time: 0 and 1. This is a concept hard to grasp at the macroscale where we live. Nevertheless, at the atomic scale, all bets are off. This fact has been proven experimentally for over 70 years. Thus, superposition allows a quantum computer to outmuscle a classical computer by performing large amounts of computation with relatively small numbers of qubits. Another mind-bender is qubit entanglement: entangled qubits transfer states, when observed, faster than the speed of light across time or space! Wrap your head around that. All in all, this chapter explores all the physical components of a quantum computer: quantum gates, types of qubits such as superconducting loops, ion traps, topological braids, and more. Furthermore, the current efforts of all major technology players in the subject are described, as well as other types of quantum computation such as quantum annealing.

Chapter 3: Behold, the Qubit Revolution

In this chapter, we look at the basic architecture of the qubit as designed by the pioneering IT companies in the field. You will also learn that although qubits are mostly experimental and difficult to build, it doesn't mean that one can't be constructed with some optical tools and some ingenuity. Even if a little crude and primitive, a quantum gate can be built using refraction crystals, photon emitters, and a simple budget. This chapter also explores superconducting loops as the de facto method for building qubits along with other popular designs and their relationship to each other.

Chapter 4: Enter IBM Quantum: A One-of-a-Kind Platform for Quantum Computing in the Cloud

In this chapter, you will get your feet wet with the IBM Q Experience. This is the first quantum computing platform in the cloud that provides real or simulated quantum devices for the rest of us. Traditionally, a real quantum device will be available only for research purposes. Not anymore, thanks to the folks at IBM who have been building this stuff for decades and graciously decided to open it up for public use.

Learn how to create a quantum circuit using the visual composer or write it down using the excellent Python SDK for the programmer within you. Then execute your circuit in the real thing, explore the results, and take the first step in your new career as a quantum programmer. IBM may have created the first quantum computing platform in the cloud, but its competitors are close behind. Expect to see new cloud platforms soon from other IT giants. Now is the time to learn.

Chapter 5: Mathematical Foundation: Time to Dust Up That Linear Algebra

Matrices, complex numbers, and tensor products are the holy trinity of quantum computing. The bizarre properties of quantum mechanics are entirely described by matrices. It is the rich interpretation of matrices and complex numbers that allows for a bigger landscape resulting in an advantage over traditional scaler-based mathematics. Quantum mechanics sounds and looks weird but at the end is just fancy linear algebra.

Chapter 6: Qiskit, Awesome SDK for Quantum Programming in Python

Qiskit stands for Quantum Information Software Kit. It is a Python SDK to write quantum programs in the cloud or a local simulator. In this chapter, you will learn how to set up the Python SDK on your PC. Next, you will learn how quantum gates are described using linear algebra to gain a deeper understanding of what goes on behind the scenes. This is the appetizer to your first quantum program, a very simple thing to familiarize you with the syntax of the Python SDK. Finally, you will run it in a real quantum device. Of course, quantum programs can also be created visually in the composer. Gain a deeper understanding of quantum gates, the basic building blocks of a quantum program. All this and more is covered in this chapter.

Chapter 7: Start Your Engines: From Quantum Random Numbers to Teleportation and Super Dense Coding

This chapter is a journey through three remarkable information-processing capabilities of quantum systems. Quantum random number generation explores the nature of quantum mechanics as a source of true randomness. You will learn how this can be achieved using very simple logic gates and the Python SDK. Next, this chapter explores two related information processing protocols: super dense coding and quantum teleportation. They have exuberant names and almost magical properties. Discover their secrets, write circuits for the composer, execute remotely using Python, and finally interpret and verify their results.

Chapter 8: Game Theory: With Quantum Mechanics, Odds Are Always in Your Favor

Here is a weird one: this chapter explores two game puzzles that show the remarkable power of quantum algorithms over their classical counterparts – the counterfeit coin puzzle and the Mermin-Peres Magic Square. In the counterfeit coin puzzle, a quantum algorithm is used to reach a quartic speed up over the classical solution for finding a fake coin using a balance scale a limited number of times. The Mermin-Peres Magic Square is an example of quantum pseudo-telepathy or the ability of players to almost read each other's minds, achieving outcomes only possible if they communicate during the game.

Chapter 9: Quantum Advantage with Deutsch-Jozsa, Bernstein-Vazirani, and Simon's Algorithms

This chapter looks at three algorithms of little practical use but important, because they were the first to show that quantum computers can solve problems significantly faster than classical ones: Deutsch-Jozsa, Bernstein-Vazirani, and Simon's algorithms. They achieve significant performance boost via massive parallelism by using the Hadamard gate to put the input in superposition. They also illustrate critical concepts such as oracles or black boxes that perform some transformation on the input, and phase kickback, a powerful technique used to alter the phase of inputs so they can cancel each other.

Chapter 10: Advanced Algorithms: Unstructured Search and Integer Factorization with Grover and Shor

This chapter showcases two algorithms that have generated excitement about the possibilities of practical quantum computation: Grover's Search, an unstructured quantum search algorithm capable of finding inputs at an average of the square root of N steps. This is much faster than the best classical solution at N/2 steps. It may not seem that much, but when talking about very large databases, this algorithm can crush it in the data center. Expect all web searches to be performed by Grover's in the future. Shor's

Integer Factorization: the notorious quantum factorization that experts say could bring current asymmetric cryptography to its knees. This is the best example of the power of quantum computation by providing exponential speedups over the best classical solution.

Chapter 11: Quantum in the Real World: Advanced Chemistry and Protein Folding

Quantum is already working hard to make a difference in the fields of Chemistry and Medicine. This chapter showcases two amazing real-life experiments that illustrate its power: ground states are important in molecular chemistry, with most elements modeled using lattices where vertices represent interacting atoms. In this chapter, you will learn how to minimize the energy Hamiltonian of a molecule to reach its ground state using lattices. Next, proteins are the fundamental building blocks that power all life. Reliably predicting protein structures is extremely complicated and can change our understanding of nature. In this experiment, you will learn about protein amino acids, peptides, chains, nomenclature, and more; and best of all, you will learn how its structure can be predicted using a quantum computer.

## CHAPTER 1

Quantum Fields: The Building Blocks of Reality

The beginning of the 20th century, more specifically 1930s Europe, witnessed the dawn of arguably one of the greatest theories in human history: quantum mechanics. After almost a century of change, this wonder of imagination has morphed and taken many directions. One of these is *quantum field theory* (QFT) which is the subject of this chapter. If you enjoy physics and wish to understand why things are the way they are, then you must get your feet wet with QFT. It has been called the most successful theory in history, riding high since the 1950s and giving rise to the standard model of particle physics. This is the modern view of how nature works at the smallest scale, being proven right time and again by countless experiments and instruments like the Large Hadron Collider (LHC). All in all, the story of how QFT came to be, and the Masters of Physics behind it, is a tale of wonder, furious rivalry but ultimate collaboration.

Our story begins in 1900 when Lord Kelvin stood in front of the British Science Royal Society and enunciated: "*There is nothing else to be discovered in physics*" – a powerful statement at the time but clearly wrong in hindsight. Perhaps, we should thank the lord for such a bold proclamation because it is statements like that that drive others to prove them wrong. This was put to the test 30 years later in Germany.

Around the 1930s, the great German physicist Max Plank (1858–1947) was working on the black-body radiation problem, more specifically in the ultraviolet catastrophe. To understand this problem, let's backtrack to the physics of how materials glow in multiple colors at different temperatures. In 1900 British physicist Lord Rayleigh derived an approximation to predict that process. To accomplish his task, Rayleigh used the so-called black body, a simple object that would absorb and emit light but not reflect it.

1

© Vladimir Silva 2024
V. Silva, *Quantum Computing by Practice*, https://doi.org/10.1007/978-1-4842-9991-3_1

Note that the term *black* doesn't mean its color is black but that it simply absorbs and emits light but *does not reflect it*, so when observed, you'll see its glow or *radiation*. Rayleigh's work is known as the Rayleigh-Jeans law for spectral radiation of a black body as a function of its wavelength λ (lambda) and its temperature in Kelvin degrees (K) (see Equation 1.1):

$$B_\lambda(T) = \frac{2cK_BT}{\lambda^4} \tag{1.1}$$

where

- c = speed of light (299792458 m/s)

- K_B, the Boltzmann constant = 1.38064852 × 10-23 m2 kg s-2 K-1

- λ = wavelength

- T = temperature in Kelvin degrees

Enter Max Planck, the Father of Quantum Mechanics

The Rayleigh-Jeans law works great for higher wavelengths (in the infrared spectrum outside of visible light) but gives infinite values in the visible spectrum. Figure 1-1 shows a graph of the Rayleigh-Jeans spectral radiance for wavelengths of visible and infrared for a black body at 5000 degrees Kelvin. This is what is known as the ultraviolet catastrophe: the infinite values of radiation of light in the visible spectrum as predicted by classical physics. This is simply not possible; if this was true, then we'll all get cooked up by simply getting close to a candle light! Max Planck realized this and found a solution in the 1930s earning him a Nobel Prize and a place in history.

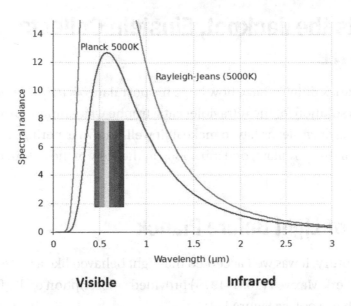

Figure 1-1. *Graph of the Rayleigh-Jeans law vs. Planck's solution for the ultraviolet catastrophe*

Planck altered Rayleigh's original derivation by changing the formula to match experimental results as shown in Equation 1.2.

$$B_\lambda(T) = \frac{2cK_BT}{\lambda^4} \tag{1.1}$$

$$B_\lambda(\lambda,T) = \frac{2hc^2}{\lambda^5} \frac{1}{e^{\frac{hc}{\lambda K_BT}} - 1} \tag{1.2}$$

where h is Planck's constant = 6.62×10^{-34} m²kg/s.

He made an incredible assumption for the time: energy can be emitted or absorbed in discrete chunks which he called *quanta*: $E = hv = h\frac{c}{\lambda}$ where v is the frequency. Note that frequency equals the speed of light divided by the wavelength $v = \frac{c}{\lambda}$. This may seem trivial nowadays, but in the 1930s was ground-breaking; not even Planck fully understood what he had unleashed. He gave birth to a brand new theory: quantum mechanics.

3

Planck Hits the Jackpot, Einstein Collects a Nobel Prize

So at the time, Planck didn't realize how huge his postulate of *energy quanta* was, as he admitted that his solution for the ultraviolet catastrophe was simply a workaround for the maths of the Rayleigh-Jeans law to make it fit well-known experimental results. To grasp the power of this postulate, one must look at the view of the nature of light pre-post Planck's era.

The Nature of Light Before Planck

Since the 19th century, it was well accepted that light behaved like a wave. Scottish physicist James Clerk Maxwell (1831–1979) provided a description of the fundamental properties of such waves (see Figure 1-2).

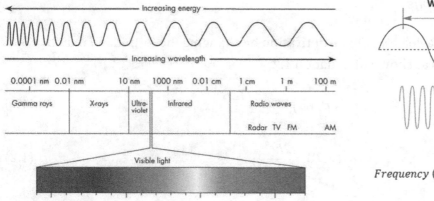
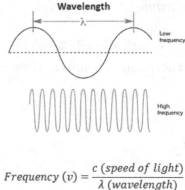

$$Frequency\ (v) = \frac{c\ (speed\ of\ light)}{\lambda\ (wavelength)}$$

Figure 1-2. *The nature of light in the 19th century*

- A fundamental property of a light wave is its wavelength or lambda (λ).

- Look at the right side of Figure 1-2: At very short wavelengths, we have lots of waves; the reverse is also true at higher wavelengths. This is the frequency (v), a second fundamental property of light waves.

It seems logical to assume that at high frequencies (short wavelengths), the energy of the wave is higher (as there is more stuff flowing in) and that at lower frequencies (higher wavelengths) the energy decreases. Therefore the energy (E) is directly proportional to its frequency (v) and inversely proportional to its wavelength (λ). This knowledge gave rise to the standard spectrum of light in the 19th century:

- On the left side of the spectrum (at the shortest wavelengths between 1 picometer and 0.01 nanometers [nm]), sit the *gamma rays*: very dangerous, the usual result of a supernova explosion, they are the most energetic. A gamma-ray burst from a supernova can destroy everything in its path: all life on Earth, for example, even the solar system. You don't want to be in the crosshairs of a gamma-ray burst!

- Next, at a wavelength of 0.01–10nm, sit the well-known x-rays: very helpful for looking inside of things: organic or inorganic, but still dangerous enough to cause cancer over persistent exposure.

- At a wavelength of 10–400 nm, we have ultraviolet light (UV): this is the radiation from the sun that gives life to our Earth but can be harmful in high doses. Lucky for us, the ozone layer on Earth keeps the levels in balance to make life possible.

- At a tiny sort after the UV range sits the visible light spectrum that allows us to enjoy everything we see in this beautiful universe.

- Next, infrared at wavelengths up to 1050 nanometers. It is used in industrial, scientific, military, law enforcement, and medical applications. In such devices as night vision goggles, heat sensors, and others.

- Finally, radio waves above the infrared range. These are used by most human technology to send all kinds of information such as audio, video, TV, radio, cell phones, you name it.

After Planck, Physics Will Never Be the Same

In the 1930s Planck turned the classical understanding of the nature of light upside down. Even though his postulate of energy quanta was dubbed lunacy by most physicists and remained unnoticed for years, it will take another giant of the century, Albert Einstein, to seize on this discovery and come up with a brand new interpretation of light. Thus, the photon was born.

This is not well known to most people, but Einstein didn't win a Nobel Prize for his masterpiece on *The Theory of Relativity*, but for his work on the quantum nature of light and the photoelectric effect. Using Planck's idea, Einstein imagined light as discrete waves (particles) which he called photons. He used this to solve a paradox in the photoelectric effect unknown at the time (see Figure 1-3).

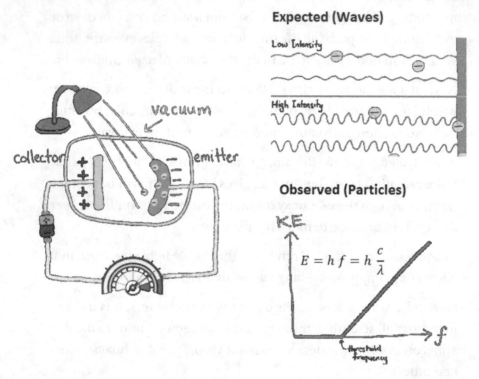

Figure 1-3. *A fresh idea on the photoelectric effect earned Einstein the Nobel Prize in Physics in 1921*

As its name indicates, the photoelectric effect seeks to describe the behavior of electrons over a metal surface when light is thrown in the mix. To this end, the experiment in Figure 1-3 was devised:

- Start with two metal plates. Let's call them the emitter and the collector. Both are attached via a cable to a battery. The negative end of the battery is connected to the emitter, and the positive to the collector.

- As we all know, electrons have a negative charge; thus, they flow to the emitter while the positive charge gathers in the collector. Remember also that opposite charges attract.

- The idea is to measure the kinetic energy of the electrons when they flow from the emitter to the collector when a light source is thrown into the emitter. To achieve this accurately, a vacuum is set among the two.

- If light flows as a wave as classical physics demands, then when the light hits the electrons, they will become energized and escape the surface of the emitter toward the collector. Furthermore, as the intensity (the amount) of light is increased, more electrons will get energized and escape in larger quantities. This increase in charge can be measured by the gauge as shown.

However, this is not what happens. Two things were observed in reality:

1. The increase in charge (the kinetic energy of the electrons) does not depend on the intensity of the light but on its frequency.

2. Even stranger, not all frequencies energize the electrons to escape the emitter. If we were to draw the kinetic energy (KE) as a function of the frequency (f) (see the lower right side of Figure 1-3), then there is a point in the curve (threshold frequency) after which the electrons escape. Values below this threshold and the electrons remain unchanged. This is a puzzle indeed!

Einstein proposed a solution to this puzzle: by postulating that energy behaves as a particle, he solved the paradox of item 2 of the list. Imagine that you are at the county fair looking to win a prize by knocking down pins with a ball. If you throw marbles at the pins, they won't budge; however, throw a baseball, and the pins will be knocked down earning you that desired prize. This is what Einstein thought occurred in this situation. Low frequency photons don't have enough energy to power up the electrons to escape the emitter. Increase the frequency of the light; it increases the energy of the electrons so they escape generating a current that can be measured. From this, a mathematical model can be derived (see Figure 1-4).

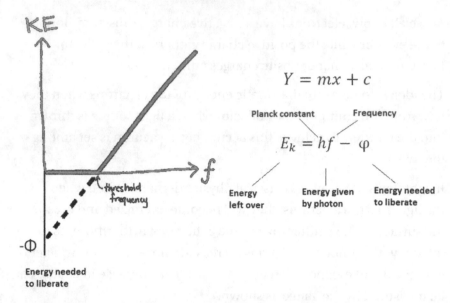

Figure 1-4. *Equation for the photoelectric effect*

Figure 1-4 shows a graph of the kinetic energy of the electron (EK) as a function of the light frequency (f). At low frequencies, no electrons escape until the threshold frequency is reached. Now, extend the line as shown by the dotted track in the figure, and we have a straight line graph (note that the point at which the dotted track intersects the Y axis is named by the Greek letter φ (Phi)). This is the energy needed to liberate the electron. Thus, this line graph can be described by the algebraic equation $Y = mx + c$ where m is the gradient and c is the Y-intercept.

Now instead of Y, substitute the kinetic energy, with the gradient m being Planck's constant (h), the frequency (f) instead of x, and c being the energy needed to liberate or $-\varphi$. Therefore, our line graph equality becomes $E_k = hf - \varphi$.

This is the equation for the photoelectric effect: the energy leftover after the electron is liberated equals the energy given by the photon minus the energy needed to liberate it.

Tip   Incidentally, the first scientist to think of light as a particle was Isaac Newton. He thought light traveled in small packets which he called co-puzzles. He also thought these packets had mass; something that is incorrect. Unfortunately, this idea never took off and lay dormant until it was revived by the Planck-Einstein revolution of the 1930s.

Quantum Mechanics Comes in Many Flavors

There is little doubt that the 1930s were the golden age of physics in the 20th century. Nobel prizes were awarded like candy, and it seemed that nothing could stop humanity in its quest to unravel the secrets of nature. Since then, quantum mechanics has stood tall for almost a century of endless theoretical and experimental challenges. All in all, it has seen a good deal of change over the years. These are the so-called interpretations of quantum mechanics, and they come in really bizarre flavors.

Copenhagen Interpretation

This is the earliest consensus about the meaning of quantum mechanics, and was born out of the golden age of physics with contributions from Max Planck, Niels Bohr, Werner Heisenberg, and others in Copenhagen during the 1920s.

The Revolution Begins with Planck, Bohr, and Schrödinger

Max Planck's postulate of energy quanta started the revolution with contributions by Einstein on the duality and/or quantum nature of light. That is, the idea that light behaves as both a wave and a particle.

Danish physicist Niels Bohr (1885–1962) funded the Institute of Theoretical Physics in Copenhagen in the 1920s to work on the brand-new field of atomic research. At the time, the atom was thought to look like a tiny solar system with a nucleus at the center made of protons, neutrons, and electrons orbiting around. This was known as the Rutherford model, but it had a terrible problem: electric charge! If the negatively charged electrons orbit around the positively charged nucleus, then as opposite charges attract, the electrons will eventually collapse into the nucleus destroying all matter in existence. Bohr foresaw this situation and used Planck's idea of energy quanta to theorize that electrons jump from one orbit to another by gaining or losing energy; something that he called a *quantum jump*. This idea later became known as the Bohr atom, but it had a weird characteristic: electrons didn't simply travel from one orbit to another. They instantaneously disappear from one orbit and reappear in another. This did not sit well with another colossus of physics: Erwin Schrödinger.

Austrian physicist Erwin Schrödinger (1887–1961) is the father of the famous wave function ψ (Cyrillic - Psi). Schrödinger was looking to describe the energy of a physical system; he came up with what is now considered the most powerful tool in physics in the last century (see Figure 1-5).

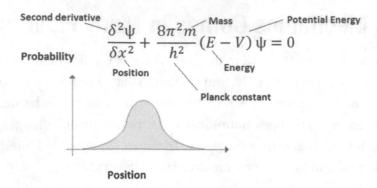

Figure 1-5. *Schrödinger wave function ψ is the cornerstone of quantum mechanics*

Schrödinger detested Bohr's interpretation of the atom famously stating that "If I am to accept the quantum jump, then I am sorry I ever got into the field of atomic research." As a matter of fact, his wave function was an attempt to defeat Planck-Bohr-Einstein. He wanted to throw away the nascent theory of energy quanta and return to the continuous classical model of wave physics, even pushing the idea that all reality can be described entirely by waves. So why is ψ used nowadays everywhere in quantum mechanics? Thank this to our next physicist: Max Born.

German-Jewish physicist Max Born (1882–1970) took Schrödinger's wave function in an entirely new direction. Born proposed a probabilistic interpretation of ψ, that is, the state of a particle exists in constant flux, and the only thing we can know is the probability of the particle at a given state. Born postulated that this probability is $P = \psi^2$. Needless to say, Schrödinger didn't like this at all as he thought his wave function was being misused. He took a swing at Born with his now famous thought experiment: the quantum cat. But before we check if the cat in the box is dead or alive and why, consider this witty story: In the quintessential American cartoon *Futurama* (by Matt Groening – creator of *The Simpsons*), our hero Fry enrolls in the police academy in New-NewYork on Earth in the year 3000. One day while on patrol, Fry chases a bandit carrying a mysterious box in the trunk of his car. Once in custody, the bandit is revealed to be Werner Heisenberg. Fry looks at the box with a face full of trepidation and asks: "What's in the box?" To which Heisenberg replies, "a cat." "Is the cat dead or alive?" asks Fry. Heisenberg replies: "the cat is neither dead nor alive but in a superposition of states with a probability assigned to each." Long story short, Heisenberg the bandit is arrested as a major violator of the laws of physics. This was a funny tale for the physics buff. Nevertheless, it shows the quantum cat has become folklore, and the prime example used to explain the probabilistic nature of quantum mechanics.

The powerful *Heisenberg's uncertainty principle (HUP)* is the work of German physicist Werner Heisenberg (1901–1976), and it is one of the foundations of quantum theory. It describes a degree of uncertainty in the relationship between the position (x) and the momentum (ρ) of a particle. More clearly, we can measure the exact *position* or *momentum* of a particle but not both. The uncertainty principle arises from the fundamental wave-matter duality of quantum objects (see Figure 1-6).

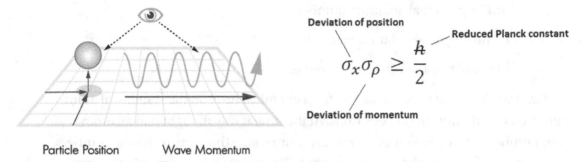

Figure 1-6. *Uncertainty is a fundamental property of the wave-particle duality of a quantum object's complementary variables such as position and momentum*

Tip   A remarkable point is that at the beginning, this degree of uncertainty was confused with the *observer effect*, which states that the act of measurement alters the state of a quantum system. As a matter of fact, Heisenberg himself used the observer effect as a physical explanation of this postulate. Since then this has been proven untrue with a rigorous mathematical derivation of HUP provided by physicist Earle Hesse Kennard in 1928.

The uncertainty principle has a profound effect in the world of thermodynamics: for example, it gave rise to the notion of *zero-point energy*. In the Kelvin scale of temperature, zero kelvin is called the *absolute zero* or the temperature at which all molecular activity stops. This fact is forbidden by quantum mechanics and the uncertainty principle because, if all molecular activity ceases, then the position and momentum of a particle will be known. This is not possible; you either know the position or the momentum of a particle but not both. Thus even at absolute zero, particles are vibrating with a tiny amount of energy, hence the term zero-point energy.

Austrian physicist Wolfgang Pauli (1900–1958) was a dear colleague of Einstein and a Nobel laureate for the remarkable *Exclusion Principle* which states that no two electrons can have the same set of quantum numbers. These numbers describe the state of the electron; therefore, no two electrons can be in the same quantum state at a time. In Pauli's time, the chemical effect of the electron was described by a set of three quantum numbers:

- n: The principal quantum number

- l: The orbital angular momentum

- m_l: The magnetic quantum number

Pauli studied experimental results from chemical tests on the stability of atoms with even vs. odd numbers of electrons. At the time it was thought that an atom with even numbers of electrons was chemically more stable than one with odd numbers. Furthermore, these numbers were thought to be arranged in symmetric clusters or closed shells around the nucleus. Pauli realized that these complex shells can be reduced to a single electron by adding a new quantum number to the trio above. Pauli introduced a new two-valued quantum number that will later be known as the *quantum spin*. Pauli's discovery was later generalized for all particles in the standard model:

- Fermions: Named after one of the architects of the nuclear age (Enrico Fermi), these obey Fermi-Dirac statistics and Pauli's exclusion principle. Fermions have a half-integer spin and include electrons, quarks, and leptons (electrons, neutrinos).

- Bosons: These obey Bose-Einstein statistics and do not follow Pauli's exclusion principle. Furthermore, they have integer value spin and include photons, gluons, W-Z bosons, and the almighty Higgs boson (the so-called god particle).

Pauli's exclusion principle is important in that it helps explain the complex shell structure of atoms and its effect on their chemical stability. It also explains the way atoms share electrons explaining the chemical variety of elements in nature and their combinations. For this, Pauli received a Nobel Prize in 1945 for "a contribution through his discovery of a new law of Nature, the exclusion principle or Pauli principle," with the incredible honor of being nominated by Albert Einstein.

The Genius of Paul Dirac

English physicist Paul Dirac (1902–1984) is considered one of the most significant contributors to the development of quantum mechanics and quantum electrodynamics. You probably heard of the term antimatter, that is, matter with the same mass as regular matter but opposite charge. Dirac was the first to derive an equation to predict its existence. Among many of Dirac's contributions are

- Dirac equation: This equation is considered an incredible achievement for quantum mechanics for two important reasons: First, it was an attempt to account for special relativity (space-time coordinates) within Schrödinger's wave function (see Figure 1-7). Such a feat is considered to be the holy grail of physics: merging relativity and quantum mechanics into a single theory of everything. Unfortunately, Dirac's equation fell a little short of the feat of the millennium. We'll explain that later on. Second, it predicts the existence of antimatter, unsuspected and unobserved at the time, yet confirmed experimentally years later via experiments of particle colliders.

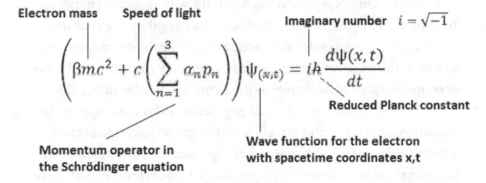

Dirac Equation using natural units (c = h = 1)

$$(i\Upsilon_u \delta^u - m)\psi = 0$$

Pauli Matrices   Derivative in 4 dimentions

Figure 1-7. *The Dirac equation was the first attempt to inject relativistic space-time in the context of quantum mechanics*

13

- Dirac's intentions of treating the atom in a manner consistent with relativity had a profound effect in the structure of matter. His equation introduced two new mathematical objects which now are fundamental in physics and quantum field theory:

  - a_k and β: These are 4x4 matrices closely related to Pauli's matrices. Remembering Pauli's exclusion principle, it introduced a new quantum number to explain electron shell clusters using a 2x2 matrix that was later known as electron spin (or spinor). In the same line, Dirac's matrices are called bispinors.

  - A four-component wave function ψ: It has four components because its evaluation at any given point is a bispinor. Physically, it is interpreted as the superposition of a spin-up electron, spin-down electron, spin-up positron, and spin-down positron. Note that Dirac's four-component wave function differs from Pauli's two-component wave function and Schrödinger's wave function for a single complex value.

- Hole theory: Dirac's equation has solutions with negative energies. To cope with this fact, Dirac introduced the hypothesis of *Hole theory*. This theory postulates that the vacuum is a many body quantum state where all the negative energy electron eigenstates are occupied. This description came to be known as the *Dirac sea*. Furthermore, since Pauli's exclusion principle forbids two electrons from occupying the same quantum state, additional electrons will be forced to occupy a positive eigenstate with positive-energy electrons forbidden from decaying into negative-energy eigenstates. Dirac reasoned that there may be unoccupied negative-energy eigenstates in this sea which he called holes reasoning that they behave like positively charged particles because positive energy is required to create a particle-hole pair from the vacuum. He initially thought that the hole may be the *proton*; however, it was pointed out later that the hole should have the same mass as the electron; thus, it could not be the proton as it is around 1800 times as massive as the electron. This hole was later identified as the *positron* which was discovered experimentally by American physicist Carl Anderson in 1932!

Einstein vs. Bohr, Nonlocality and Spooky Action at a Distance (EPR Paradox)

In the early part of the 19th century, an unknown battle was being fought by two titans of physics: Niels Bohr and Albert Einstein. Einstein did not like Max Born's probabilistic interpretation of the wave function. He wanted to extend his relativity to the atomic scale for a single unified theory of physics. Thus in 1935, he along with colleagues Boris Podolsky and Nathan Rosen published the notorious EPR paradox. The goal of this paper was to drive a *coup de grâce* at the heart of quantum mechanics by showing the absurdity of one of its fundamental principles: entanglement. Entanglement is a fundamental property of quantum systems that originates when two particles interact with each other. For example, if one has spin-up, the other particle will instantaneously show spin-down when measured, and thus they are said to be entangled. The bizarre part is that this event occurs instantaneously across space, even time (nonlocality). So, for example, take two entangled particles, leave one on Earth, and move the other to the edge of the solar system, then perform a measurement on the spin of the first. The second particle at the edge of the solar system will instantaneously take the opposite spin value. This seemed absurd to Einstein who believed that the speed of light was the ultimate speed limit in the universe. If nothing can travel faster than light, how can the first particle notify the other about its spin instantaneously? Einstein called this *spooky action at a distance*.

Tip   Einstein abhorred the probabilistic interpretation of quantum mechanics because he could not bear the idea that the act of observation (measurement) is what defines the state of a particle. He believed that states (properties) of a particle were defined at the moment of its creation, famously writing to Bohr, "*God does not throw dice.*" To which Bohr replied: "*You should stop telling God what to do.*" Einstein sought to defeat this idea, and he spent the last decades of his life looking for the holy grail of physics: a unified theory of relativity and quantum mechanics. He was unsuccessful, and so the holy grail still eludes us: the mother of all equations to unite the Heavens, the Earth, and the atom.

Bell's Theorem Settles Einstein vs. Bohr and the EPR Paradox

Einstein vs. Bohr raged over the years with no clear winner in sight. Both physicists passed away without settling their differences. However, in 1964, Irish physicist John Stewart Bell (1928–1990) published a theorem to settle things once and for all. Bell's theorem did not seek to prove who's right: quantum mechanics or relativity. It simply provides the means to test the principle of nonlocality in entangled particles. In simple terms, Bell's theorem states that the sum of probabilities for a correlated three variable quantum system is less than or equal to 1. That is, $P(A = B) - P(A = C) - P(B = C) \leq 1$.

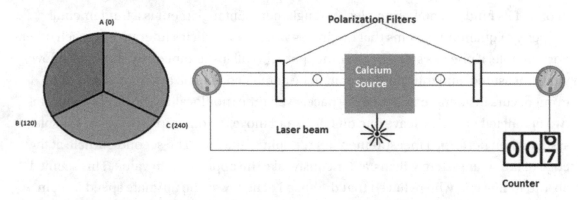

Count	A(0)	B(120)	C(240)	[AB]	[BC]	[AC]	Sum	Average
1	A+	B+	C+	1(++)	1(++)	1(++)	3	1
2	A+	B+	C-	1(++)	0	0	1	1/3
3	A+	B-	C+	0	0	1(++)	1	1/3
4	A+	B-	C-	0	1(--)	0	1	1/3
5	A-	B+	C+	0	1(++)	0	1	1/3
6	A-	B+	C-	0	0	1(--)	1	1/3
7	A-	B-	C+	1(--)	0	0	1	1/3
8	A-	B-	C-	1(--)	1(--)	1(--)	3	1

Figure 1-8. *Explanation of Bell's theorem using photon polarization at three angles*

To illustrate Bell's theorem, consider photon polarization at three different angles (A, B, and C) in Figure 1-8. We seek to calculate the minimum probability that the polarization (indicated by +/–) for two neighbors is the same. For that purpose, we use a table with the eight possible permutations of A, B, and C plus the neighbor polarization (+/– columns 5,6,7). We also calculate the sum and average. Note that equal neighbor polarization is indicated by ++ or –– in which case we count a 1. The average is the sum divided by 3. The results above show that the minimum probability must be greater than or equal to 1/3; what Bell's theorem is saying here is that if reality is defined by the act of observation as quantum mechanics predicts, then the minimum probability must be less than 1/3.

On the other hand, if relativity is correct (the state of a particle is defined at creation), then the probability must be greater than or equal to 1/3. Now the trick is to find out if quantum mechanics violates Bell's inequality. If it does, then our universe is bizarre (Bohr and Planck were correct and quantum mechanics is saved). On the other hand, if the inequality is not violated, then Einstein's relativity wins and quantum mechanics is wrong.

Amazingly, in 1982, French physicist Alain Aspect (1947–present) came up with an experiment to test if quantum mechanics violates Bell's theorem (see top right side of Figure 1-8). The experiment used a laser and a calcium source to create pairs of entangled photons. These photons travel in opposite directions passing through a polarization filter with the results accounted for at the end. The goal was to calculate the probability that both photons either pass thru or not at different angles, count the totals, and see if the sum of probabilities is greater than or equal to 1/3. Remember that each photon pair is entangled thus *spooky action at a distance* may occur when they pass the filter. The results were astounding. The probability sum was around 1/4; Bell's inequality was violated as quantum mechanics predicted. It looks like *God does throw dice* after all!

Tip   The late physicist Steven Hawking (1942–2018) once said "Not only god throws dice but he is a compulsive gambler." Some predictions of quantum mechanics are so bizarre that they escape understanding. Our brains are wired to make sense of the world around us; however, at the quantum scale, some things cannot be understood, only accepted.

Aspect's experiment also took a shot at *spooky action at a distance*. Testing if a photon is capable of telling its entangled partner about its polarization instantaneously is not an easy task. To achieve this, the experiment was slightly modified to use an optical switch that selectively shifts optical signals on or off at a rate of around 2 nanoseconds (ns). Now it takes light traveling at 186 thousand miles/sec around 20 ns to travel from one side of the experiment to the other (close to 14m). The goal was to run the experiment again and see if the new results match the old ones. If they do then, the photon was able to tell its partner about the polarization faster than light can travel from one end of the experiment to the other, something that is forbidden by relativity which says that nothing in the universe can travel faster than the speed of light. The results matched exactly in both instances: astounding and scary at the same time; imagine if both photons were placed at the edges of the galaxy.

In an interview for the British BBC, John Bell spoke about these results: In physics, some things escape our understanding, and we are left with no choice but to accept them, as if reality is playing a trick on all of us. Amazing things occur at the quantum scale, but the ultimate irony is that we cannot use them. Even though entangled particles can send signals faster than light across huge distances, information cannot be sent in the signal. A fact that is also predicted by quantum mechanics; what a bummer, forget about texting your alien buddies on Alpha Centauri.

Tip   It is hard to believe that Bell's masterful theorem remained unnoticed for years. Of course, that changed to the point that prominent physicists have called it "The most profound discovery in science."

Consciousness, Mysticism, and the Collapse of the Wave Function

Quantum mechanics tells us that the states of particles exist in superposition within a probabilistic curve or wave function. Furthermore, when a measurement is performed on the particle via a detector or measuring device, the wave function is said to collapse to a single state. Why the collapse remains a mystery, all we know is that the collapse signifies the transition from the quantum to the classical realm. Nevertheless, this seems to occur whenever a quantum system interacts with the outside world. When such bizarre physical phenomena were discovered, physicists turned to philosophy and mysticism to make sense of what is going on. As a matter of fact, sacred Hindu texts like The Vedas, Bhagavad Gita, Ramayana, and Mahabharata have entities popping in and out of existence. In the sacred text of the Rigveda, the speed of light is calculated at 2202 Yojanas in a half Nimesa which in ancient units translates to around 185K miles/second.

Quantum principles have been used through the ages to validate metaphysical concepts such as divinity, consciousness, and positive thinking among others. But setting all this aside, why has quantum mechanics convinced leading physicists of the intrinsic role mind-consciousness could play in reality? There is a fundamental principle in science called causality. It is so embedded in our everyday existence that we take it for granted: cause-and-effect, for every action there is an equal opposite reaction, actions have consequences, etc. But what if there was scientific proof that causality can be violated? Such proof will be incomprehensible as it would challenge the very fabric of reality; after all, a scientist's brain is wired to make sense of physical phenomena via observation. Such an experiment does exist, and it is called the double-slit experiment first performed by Thomas Young in 1801 (see Figure 1-9).

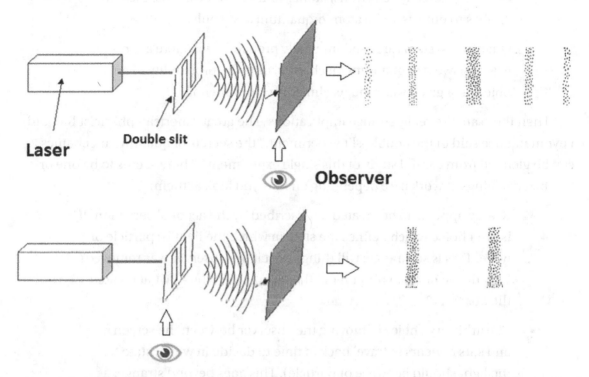

Figure 1-9. *The double-slit experiment showing wave-particle duality appears to violate causality*

In the double-slit experiment in Figure 1-9, photons originating from a laser beam travel through a metal plate with two close slits with the resulting projection recorded on a detector screen. In the first step, an observer or measuring device is placed on the screen resulting in the interference pattern shown on top of the figure. In this case, the pattern indicates that light behaves like a wave. Now, if the observer is moved close to the two slits to look at which slit the photon goes through, then the resulting pattern changes to a cluster of two lines (bottom of the figure); here the photons behave like particles. Why this happens is one of the great mysteries in physics. A few concrete conclusions were drawn by experts at this point:

- The experiment illustrates the wave-particle duality of photons intrinsic to all quantum phenomena. Note that since the days Thomas Young first performed the experiment, the same results have been replicated for electrons, atoms, and molecules; thus, the effect applies to objects at the atomic-quantum scale only.

- Quantum systems are fundamentally probabilistic: There are no exact answers in quantum mechanics, only the probability of an object at a given state and its almighty wave function.

Then there are the really strange implications. The great American physicist Richard Feynman once said of the double-slit experiment, "the secrets of quantum mechanics can be gleaned from careful study of this single experiment." There seems to be one or two bizarre things at work here depending on how you look at them:

- Reality appears to be created or described by the act of observation. It is our choice which defines the state in which the light is: particle or wave. This is so strange; will things exist if the observer was removed? Why do we need an observer in the first place? Who or what created the observer?

- The arbitrary choice of moving the observer between the screen and slits appears to travel back in time to decide in which state the light should be (wave or particle). This goes beyond strange; a more palatable explanation for this would be to think in terms of superposition of states: the light exists in probabilistic superposition of wave-particle states when not observed.

So the role consciousness plays in this saga remains. Is consciousness-observation fundamental to our reality? If so, wouldn't that challenge scientific dogma which says that consciousness is an isolated illusion of the brain? Science allows us to gather independent and verifiable evidence untainted by expectations and flourishes by removing all traces of the mind from physical phenomena.

Science seeks to build objective and reliable models of the world thriving on the standard and verifiable experimental method. Quantum mechanics now seems to reveal that, at the deepest level of nature, objective science is no longer possible. Could the mind play an intrinsic role in the unfolding of the world? Max Planck once said: "*It seems like the mysteries of nature cannot be solved by ourselves as we are part of nature and thus the mystery we are trying to solve in the first place.*"

Tip   Some quantum mechanics principles appear to defy understanding, and when faced with the unbelievable, master physicists like Planck, Schrödinger, and others have turned to philosophy to make sense of their scientific discoveries.

Many Worlds Interpretation

In the many worlds interpretation (MWI) of quantum mechanics, the wave function is universal. It exists in all realities, and it does not collapse in our universe when we observe phenomena. The multiverse is much weirder than we ever imagined.

In the Copenhagen interpretation, the collapse of the wave function signals the transition between the quantum and classical realms. However, there is another way to interpret this, which is in a set of infinite realities where all outcomes are possible.

When Erwin Schrödinger came up with his quantum cat thought experiment to challenge the principle of superposition that he called absurd, he thought that if we open the box and find out the cat is alive is because we are part of an entire quantum timeline in which the poisoning of the cat never occurred. Nonetheless, there is an equally valid timeline in which the cat died with another version of us experiencing it. Not only that, but the number of timelines is infinite and occurs simultaneously. Sounds outrageous, but it is a very serious interpretation of the mathematics.

Tip   The many worlds interpretation was proposed by American physicist Hugh Everett (1930–1982) in his 1957 PhD thesis "Theory of the Universal Wave Function."

In the context of the double-slit experiment, the Copenhagen interpretation says that, at the moment of observation, all superimposed photon trajectories (histories) merge (collapse) into a single timeline of the observer's reality. The many worlds interpretation on the other hand says that this merge never happens, all possible histories continue, and we find ourselves in one of those timelines (see Figure 1-10). Note that all histories are equally likely. However, some are similar to each other, and we tend to land in the most common history.

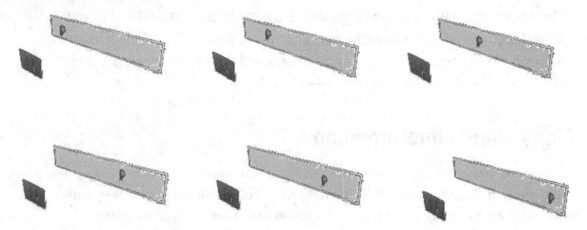

Figure 1-10. *In MWI all possible photon trajectories continue in their own timelines*

Tip   Everett's idea was not taken seriously when first proposed, probably because he was a graduate student at the time.

All in all, MWI may be the purest interpretation of quantum mechanics as its maths does not require the collapse of the wave function. Nevertheless, the idea of realities branching out of each other at every instance induces an existential crisis that could justify its unpopularity. Imagine infinite versions of yourself out there going through every possible life path. It just sounds too bizarre. All in all, the most important specifics about MWI are

- Many worlds is a deterministic interpretation: It eliminates the probabilities intrinsic to the wave function by postulating that any given timeline is a predictable chain of cause and effect. No superposition of states.

- Even though it has gained popularity in recent years, there is no evidence of its existence. It is supported by the mathematics of quantum mechanics but still remains an interpretation.

- It has not provided predictions to distinguish itself from other interpretations, nor is it complete in its explanation. It fails to explain what happens when neighboring histories interact or why the wave function translates to probabilities.

- Aside from the existential crisis it incites, it presents another deep philosophical unease: What happens to free will? If we live in a deterministic universe, then we are making all possible decisions at any given time; therefore, we have no free will. A valid thesis yet a dreadful view of existence.

Supplementary Interpretations

When it comes to interpretations of quantum mechanics, Copenhagen and many worlds are the most popular. Yet there are many more, which fall in two big categories: collapse vs. no-collapse of the wave function, further classified by determinism, nonlocality (spooky action at a distance), and observer presence. The following list describes some of them according to the collapse of the wave function.

Conscious Observer

Also known as the von Neumann-Wigner interpretation, proposed by Hungarian mathematician John von Neumann (1903–1957) with contributions by Hungarian physicist Eugene Wigner (1902–1995). It dwells in the realm of philosophy by hypothesizing that the wave function is universal and that it is the consciousness of the experimenter which collapses it. There is not much meat in the bones of this theory; nevertheless, over the years it has branched into ideas such as

- Subjective reduction: As consciousness collapses the wave function, there is a point of intersection between quantum mechanics and mind-body. Researchers in this field are hard at work on finding a correlation between conscious and physical events.

- Participatory anthropic principle: Championed by legendary American physicist John Archibald Wheeler (1911–2008). It says that consciousness plays a role in bringing the universe into existence. A bizarre idea, if things can get any weirder; all in all, consider this: if consciousness may bring reality into existence, then what happened before humans evolved to try to uncover it? Could a dinosaur's consciousness do the trick? Furthermore, if consciousness is an intrinsic property of life, what happened before life evolved on Earth? Did reality exist at the dawn of the solar system or after the big bang?

Quantum Information

A truly fascinating theory, this is an attempt to eliminate the indeterminism (chance) in the collapse of the wave function by taking cues from standard quantum physics which says that information is recorded irreversibly. In this context, information means a quantity that can be understood mathematically and physically, and irreversible means that quantum states cannot roll back to previous ones due to the second law of thermodynamics: entropy. Entropy states that the disorder (entropy) of the universe can only increase. Perhaps, this could be better understood from the perspective of the so-called arrow of time. That is, because time can only increase, and because of Heisenberg's uncertainty principle where quantum states are intrinsically random, recorded information is irreversible due to the fundamental uncertainty of quantum mechanics. All in all, the quantum information interpretation is based on three principles:

- The wave function evolves deterministically, going through all possibilities. A particle will randomly choose one of those possibilities to become real.

- The conscious observer is allowed; however, it cannot gain knowledge until information has been recorded irreversibly in the universe. Once recorded, the information becomes knowledge in the observer's mind.

- The measuring apparatus is quantum not classical, but it can be statistically determined and capable of recording irreversible information. So is the human mind that gains knowledge.

Tip   In this interpretation, there is only one world: the quantum world, and the quantum to classical transition. Furthermore, the determinism of classical laws of motion and causality are fundamentally statistical. Everything is probabilistic, but near certainty.

The following theories eradicate the notion of the wave function collapse.

Pilot Wave Theory

Also known as *de Broglie-Bohm* theory named after pioneer physicists Louis de Broglie (1892–1987) and David Bohm (1917–1992). It accommodates the wave function with the notion of *configurations* (the position of all particles in a quantum system). This theory is *deterministic* (no randomness is allowed); this implies that configurations exist even when systems are not observed (no superposition of states). It is also *nonlocal* and accepts spooky (instantaneous) action at a distance. Most notable is the presence of a *guiding equation* that governs the evolution of the configurations over time. In particular, this theory consists of two components:

- A configuration for the entire universe. These are the positions of all particles in our universe $q(t) \in Q$ where Q is the configuration space.

- A pilot wave $\psi\,(q, t) \in C$. This is a two-component wave function that governs the evolution of the configuration over time (t).

So, at every moment there exists not only a wave function (pilot wave), but also a well-defined configuration of the whole universe. This effectively gets rid of the indeterminism of the Copenhagen interpretation and the superposition of states (no quantum cat is allowed). Thus, what we perceive as reality is made by the identification of the configuration of our brain with some part of the configuration of the whole universe.

- Double-slit: In the context of the double-slit experiment, pilot wave says that each photon has a well-defined trajectory that passes exactly through one of the slits. This is in contrast to Copenhagen which states that the photons are not localized in space until detected (observed). Furthermore, the final position of the particle on the detector screen and the slit through which it passes is determined by the initial position of the particle; with the crucial assumption that the initial position is not knowable or controllable by the experimenter, so there is an appearance of randomness in the pattern of detection. According to Bohm, the wave function interferes with itself and guides the particles through the quantum potential in such a way that the particles avoid the regions in which the interference is destructive and are attracted to the regions in which the interference is constructive resulting in the pattern obtained experimentally.

- Relativity: Pilot wave conflicts with *special relativity* in the sense that it is *nonlocal* (accepts instantaneous action at a distance). Over the years, several extensions have been added in an attempt to overcome this conflict. Bohm himself in 1953 presented an extension to the theory using absolute time (where time is the same everywhere – something that is a big no in special relativity; where time is highly malleable and may be different relative to the position of the observer).

- Heisenberg's uncertainty principle: Copenhagen tells us that we cannot measure two correlated variables (position and momentum) in a quantum system at the same time due to the intrinsic uncertainty of quantum mechanics. In pilot wave theory, however, we can measure the position and momentum of a particle at the same time. Each particle has a well-defined trajectory, as well as a wave function. Observers have limited knowledge as to what this trajectory is (and thus the position and momentum). It is the lack of knowledge of the particle's trajectory that accounts for the uncertainty.

- Entanglement and Bell's theorem: Pilot wave theory makes the same empirically correct predictions for the Bell test experiments as ordinary quantum mechanics. This is because of the fundamental *nonlocality* of this theory. As stated in the previous section (see Bell's theorem), in 1982 Alain Aspect showed experimentally that Bell's inequality is violated; furthermore, he showed the instantaneous (faster than light) action between the two entangled photons as predicted. Pilot wave theory describes the physics of Aspect's experiment by setting up a wave equation for both particles with the orientation of the apparatus affecting the wave function. It is the wave function that carries the faster-than-light effect when changing the orientation of the polarization filters. It is worth reiterating that quantum mechanics as well as Bell's theorem and pilot wave theory do not allow *information* to travel faster than light. A very important distinction.

Time-Symmetric Theories

These are a set of theories championed by pioneers of quantum field theory Richard Feynman and John Wheeler. They introduce the concept of *retro-causality*. This is the notion that events in the future can affect ones in the past (just like events in the past affect ones in the future). In time symmetry, a single measurement cannot describe the state of a particle, but given two measurements performed at different times, it is possible to calculate the exact state of the system at all intermediate times. This notion affects the fundamentals of quantum systems in two ways:

- The collapse of the wave function is not a physical change to the system, but a change in our knowledge of it due to the second measurement.

- When it comes to entanglement, it is not a true physical state but just an illusion created by ignoring retro-causality. The point where two particles become entangled is simply a point where each particle is being influenced by events that occur to the other particle in the future.

De-coherence

This is an interpretation introduced in 1970 by the German physicist Heinz Dieter Zeh (1932–2018). According to Zeh, de-coherence is the loss of information from a system into the environment. For this to occur the following must be true:

- Viewed in isolation, the system's dynamics are non-unitary (this means that the time evolution of a quantum state according to the Schrödinger equation is represented by a non-unitary operator). Although the combination of the isolated system and environment evolves in a unitary fashion.

- The dynamics of the system alone must be irreversible.

- Entanglements can be generated between the system and environment having the effect of transferring quantum information to the surroundings.

Even though de-coherence discards the collapse of the wave function, it is an attempt to understand it. It does not generate an actual wave function collapse. It only provides an explanation for its apparent collapse, as the quantum system leaks information into the environment. Note that de-coherence fails to explain the *measurement problem*, that is, how and why the wave function collapses. It tries to, nonetheless, by the transition of states from the quantum system to the environment.

From Quantum Mechanics to Quantum Fields: Evolution or Revolution

After the golden age of physics in the 1930s, came what has been called the Second Quantization Revolution spearheaded by the great British physicist Paul Dirac. He made a profound contribution by finding a way to describe the behavior of the electron in relativistic terms using his great equation. Dirac not only gave us a relativistic description of the electron but also provided the basic recipe to quantize other properties of our universe such as mass, charge, position, and energy. In the following years, these ideas became the foundation to quantize all the subatomic forces in nature such as the weak, strong nuclear forces, and electromagnetism. These are examples of what is now called *quantum field theory (QFT)*: the basis of modern physics.

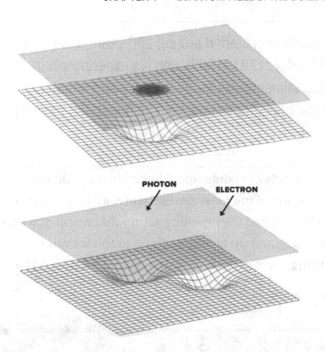

Figure 1-11. *Basic description of an electron quantum field (top) and the electron emitting a photon (bottom)*

The fundamentals of QFT are simple, yet as bizarre as the theory it splinters from:

- In the subatomic world, fields span the entire universe. There is a field for each particle in the standard model: electrons, protons, quarks, photons, etc.

- The particles we know and love are just localized vibrations in their respective field.

- Fields can interact with one another to explain how particles are created and destroyed.

To understand interactions between particles, consider Figure 1-11 showing an electron field. The electron is a localized vibration in this field (top); if the electron emits a photon, then QFT says that some of the energy of the electron sets up another vibration in the photon field (bottom) which then moves away.

Tip   The idea of a field was started around 250 years ago by the English scientist Michael Faraday (1791–1867) with his experiments in electromagnetism. Over a decade of experimentation, Faraday developed an intuition of how electric and magnetic phenomena behave. The result: the modern view of the electromagnetic field.

For a scientist like Faraday to come up with the abstract idea of a field spanning all of space, in the 1700s, was one of the most revolutionary ideas in science. Faraday's eureka moment came when he realized the repelling action that resulted when trying to put two magnets together. He imagined there is something in there even though we cannot see it. He called this something, *lines of force*; we call it the magnetic field (see Figure 1-12).

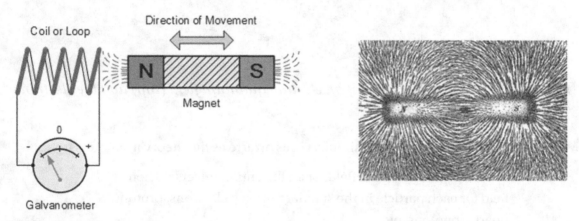

Figure 1-12. *Faraday's experiment in electromagnetic induction and his theory of lines of force gave us the present notion of the electromagnetic field*

It will take another 50 years for Faraday's idea to be scientifically proved by his pupil James Clerk Maxwell and later Heinrich Rudolf Hertz. Furthermore, another 150 years for it to be used as the basis of QFT in modern day physics.

Tip   QFT takes ideas from quantum mechanics, such as energy flowing as discrete chunks (quanta), and combines it with Faraday's idea of fields which are continuous, smooth objects oscillating in space. This is Faraday's legacy: there are no particles in our universe, only fields and the vibrations they produce. Thus, QFT seems like a simple combination of two old principles into a modern description of reality; nevertheless, its mathematics are complicated. So much so that solving most of the equations in QFT requires supercomputing power.

We Are All Made of Quantum Fields, but We Don't Understand Them

In modern physics, the periodic table describes the basic particles we are all made of. It starts with three particles: the electron (discovered by JJ Thomson in 1897) and two quarks (up and down at the top of Figure 1-13). In the 1970s, we learned that the familiar parts of the nucleus, the proton and neutron, are themselves made of smaller bits quarks, with the proton made of two up quarks and one down quark and the neutron made of two down quarks and an up quark.

Tip   The names of the quarks seem bizarre (up, down, strange, charm). Physicists in the 1970s used these names for no good reason. It's not like the up quark points upward or the down quark points downward; they are simply names given by the individuals who discovered them.

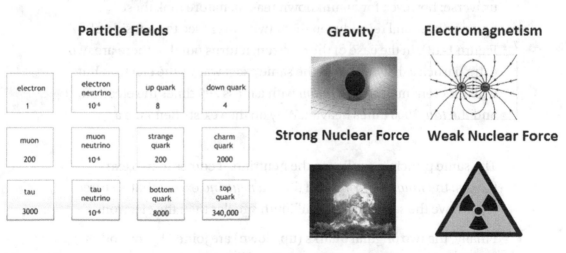

Figure 1-13.   *The physics equivalent to the periodic table showing the basic building blocks of the universe: the electron, neutrino, quarks, and the fundamental forces of nature*

It is perplexing to imagine that all that exists can be made of different arrangements of an electron and two quarks. It is a very simple and comforting picture but unfortunately incomplete.

The Recipe to Build a Universe

The periodic table of physics starts with the electron; the first and smallest particle discovered and assigned a mass of one. The mass of all other particles in the standard model is measured in terms of the mass of the electron.

- Besides the electron and up-down quarks that make up protons and neutrons, a fourth particle, the neutrino (proposed by Pauli in 1930 to explain radioactive decay), was detected experimentally in 1956. The neutrino is tiny, with a millionth the mass of the electron; it is not part of the atoms we are made of but emanates from solar radiation and is very hard to detect due to its tiny size. As a matter of fact, around 10^{14} neutrinos emanating from the sun are going through your body every minute. They don't interact with matter much, and if we wish to build a neutrino observatory, we must do it deep inside the Earth.

- Thus we have four particles that constitute the bedrock of our universe; however, for an unknown reason, nature took these four particles and reproduced them twice over (see the left side of Figure 1-13). In the case of the electron, it turns out that there are two extra particles that behave in the same exact way as the electron but with different masses: the *muon* with a mass 200 times the electron, and the *tau*, 3000 times heavier. Why do they exist? Remains a mystery.

- The same principle applies to the neutrino. It comes in two extra flavors: the *muon neutrino* and the *tau neutrino* except that all three types have the same mass, 1-millionth smaller than the electrons.

- Finally, the two original quarks (up, down) are joined by four others: strange, charm, bottom, and top, much more massive than their archetypes.

Tip   Physicists have a very clear understanding of the four cornerstone particles (electron, neutrino, up-down quarks), their properties, and roles in nature; however, why mother nature decided to reproduce them twice with different mass is unknown, and the subject of active research in particle physics. There may be more replicas out there hidden from our current collider technology.

QFT says everything is made of the 12 particle fields above, and these fields may interact not only with each other but with the four fundamental forces of nature (see Figure 1-13), which are also described by quantum fields:

- Two of these forces are very familiar: Gravity and electromagnetism. They work at the galactic scale, and we feel them every day binding the Heavens and the Earth.

- At the atomic scale there are two extra forces in action: The strong nuclear force that holds the quarks that make protons and neutrons together and the weak nuclear force which is responsible for radioactive decay and the stars in the sky.

Tip   In QFT, the four forces in nature have their own field: Faraday discovered the *electromagnetic field*. The field associated with the strong nuclear force is called the *gluon field*; the field associated with the weak nuclear force is called the *w-z boson field*.

There is a field associated with gravity too, described by Einstein as the fabric of space-time. This is the foundation of his theory of relativity and the subject of much discussion. You probably have heard that relativity and quantum mechanics are like oil and water; they don't mix at all. But that is the subject of another story.

The Fantastic Four Forces of Nature: Enter the Higgs Field

In the universe we live in, everything that exists results from the combination and interaction of all these fields; and the theory that comprises it all, the *standard model*. It has a mundane name, but it is the most successful theory in science. But, there is one extra field that became famous in recent years. The Higgs field was postulated by

Scottish scientist Peter Higgs (1929–present) in the 1960s; you may know it by its *nom de guerre*, the god particle. It is called that because the Higgs field is believed to be responsible for the mass of all particles in the standard model. By the 1970s it became an integral part of QFT, though experimental proof of its existence remained elusive until 2015 when ripples of the Higgs field were detected by the Large Hadron Collider (LHC) in Switzerland (see Figure 1-14).

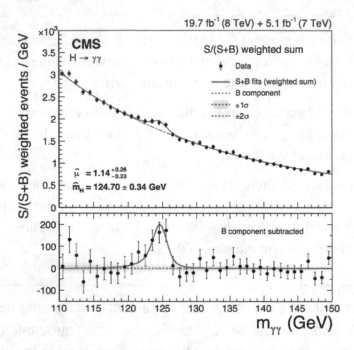

Figure 1-14. *Experimental results[1] from LHC showing a crest around 125 Giga-Electron-Volts (GeV). The crest is believed to be the so-called god particle or Higgs boson*

The Higgs boson is about 2.4×10^6 heavier than the electron and doesn't last for long, about 10^{-22} seconds. Higgs is the final piece of the puzzle, and it is important for two reasons:

- It is responsible for all the mass in the universe. It explains the property we call particle mass as a manifestation of the interaction of that particle's field with the Higgs field. Higgs gives us a clear understanding of the meaning of mass.

---

[1] Image and results courtesy of the LHC-CMS available online at https://cms.cern/news/cms-closes-major-chapter-higgs-measurements

- It is the final piece of the jigsaw: Since the 1970s the standard model has been incredibly successful in describing mother nature at the atomic scale both at the theoretical and experimental level. Physicists have been chasing the god particle since the 1970s and finally found it in 2015; but it took the most expensive scientific instrument ever built to do it: a 27 km superconducting ring and 46m long detectors at an estimated cost of USD 4.4 billion.

Standard Model and the Super-Equation of Physics

The current understanding of physics as described by the standard model has given us a *master equation* that has been dubbed the *theory of almost everything*. That is an equation that describes the behavior of the universe and everything on it (see Figure 1-15).

$$\mathcal{L} = -\frac{1}{4} F_{\mu\nu} F^{\mu\nu}$$

$$+ i \bar{\psi} \slashed{D} \psi + h.c.$$

$$+ \bar{\psi}_i y_{ij} \psi_j \phi + h.c.$$

$$+ |D_\mu \phi|^2 - V(\phi)$$

Figure 1-15. *Lagrangian of the standard model of physics as displayed in coffee mugs at CERN (image courtesy of*[2]*)*

---

[2] Sit down for coffee with the Standard Model. Available online at https://home.cern/news/news/cern/sit-down-coffee-standard-model

The so-called equation of almost everything is called a *Lagrangian* or an equation to determine the state of a changing system and the maximum possible energy it can maintain. This super equation encompasses everything we have seen so far: from Faraday's electromagnetic field, the 12 particle fields, the 4 fields for the forces of nature, and the Higgs field:

- The first line describes the forces of nature: electromagnetism and strong and weak nuclear forces.

- The second line describes how these forces act on the fundamental particles of matter, namely, the quarks and leptons.

- The third line describes how these particles obtain their masses from the Higgs boson.

- The last line enables the Higgs boson to do its job.

It looks like gibberish to the average man, but this equation correctly predicts the result of every experiment done in science. It is the pinnacle of the reductionist approach to physics. An astonishing achievement and our current limit of knowledge – the best we can do so far.

Chasing the Unexplained

In the cosmic game of science, every time you figure out a mystery, another one pops out even more mysterious. The standard model super-equation may explain everything that happens at the atomic scale; however, when looking at the sky, some things remain unexplained:

- While studying the movement of galaxies and the gravity that holds them together, astronomers found a discrepancy. Galaxies should be moving away much faster than observed for the amount of matter they contain, something is holding them, something that is invisible (it doesn't show on instruments and it doesn't interact with light): *dark matter*. Note that the name *dark* doesn't mean the matter is actually dark, but we simply don't know what it is.

- Dark matter may not be observable, but its gravitational effect can be precisely measured and it surrounds all galaxies (at around 27% of all mass energy – see Figure 1.16). Additionally, there is something even more mysterious called *dark energy*. It is believed to be culpable for the acceleration of the universe. In 1998 astronomers used the light of supernovae to calculate the rate of expansion of the universe; what they found is astounding: the universe is actually picking up speed, and its ultimate fate is a topic of hot discussion. Some say it will rip itself apart accelerating faster than light (Big Rip), others say it will disintegrate into isolated stable particles such as electrons and neutrinos, with all complex structures dissipating away (heat death), while others say it will eventually stop and gravity will take over resulting in a Big Crunch. All in all, dark energy makes up an unbelievable 68% of all there is (see Figure 1-16).

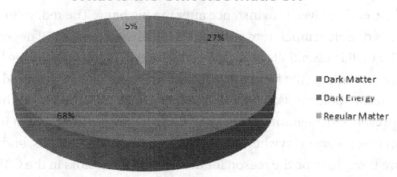

Figure 1-16. *Two mysterious things: dark matter and dark energy make up 95% of the entire visible universe*

- Inflation: Yet another enigma about the early life of the universe. Inflation is the rapid phase of expansion of the early universe discovered using the cosmic background radiation (CBR), and it is believed to be the reason why galaxies, stars, planets, and ultimately us humans exist the way we are. We know it happened, but it is not explained by the standard model equation.

Cosmic Background Radiation   **Quantum Fluctuation**

Figure 1-17. *CBR image taken by the Wilkinson Microwave Anisotropy Probe (WMAP) (left); a 3D simulation of a quantum fluctuation of the vacuum of space (right)*

In Figure 1-17 we see an image taken by NASA's WMAP instrument which depicts a snapshot of the early universe an instance after the big bang. The red, yellow, green, and blue regions denote temperature variations with an average of 2.72 degrees Kelvin (-273.15 C). The CBR is essentially an image of the fireball created by the big bang, and the temperature variations (the flickering of colors in the image) are believed to be caused by quantum fluctuations of the vacuum (right side of Figure 1-17). The reason for this is the presence of quantum fields; even though particles didn't exist in that early time, quantum fields were everywhere along with quantum fluctuations, and these fluctuations are thought to be the reason for the flickering of colors in the CBR. Because inflation occurred very quickly, it caught these fluctuations in the act and stretched them across the sky. Thus, the quantum fluctuations on the right of Figure 1-17 are the ripples of color in the CBR on the left. As a matter of fact, astronomers have done the calculations and determined that the quantum fluctuations occurred around 10^{-20} seconds after the big bang, and now they are stretched 14 billion years across the entire universe. Galaxies, stars, and everything we see exist thanks to the presence of quantum fields and their fluctuations at the big bang. Another great triumph of quantum field theory.

Yet still there are unanswered questions: for example, what field is depicted in the CBR snapshot? Some believe it is the Higgs, but others say it is not the Higgs but something new; the truth is we don't know.

Dark Energy Will Determine the Ultimate Fate of the Universe

With the information provided by the CBR and the discovery of the acceleration of the universe due to the presence of dark energy, the fate of the universe has been propelled to the top of the list of unsolved mysteries in cosmology. For this purpose, astronomers have created the *equation of state (w)*. $w = \dfrac{p}{\rho}$ where p is the pressure that dark energy puts on the universe and ρ (Greek – Rho) is the energy density (of dark energy in this case). The value of w will be the supreme arbiter of the end of days:

- A value of $w < -1$ indicates an accelerating universe, expanding at an ever-increasing rate. This expansion will continue unimpeded until it surpasses the speed of light, overcoming even the strong and weak nuclear forces that hold atoms together. Everything will rip itself apart faster than light, even quarks until nothing remains. This hypothesis was proposed by Robert R. Caldwell of Dartmouth College which estimated the Big Rip to occur in 22 billion years given $w = -1.5$. If the universe is to be no more, let it be in a blaze of glory (compared to the other alternatives).

- A value of $w = -1$ indicates dark energy is a positive cosmological constant, where the universe will continue expanding forever, and a heat death is expected to occur. The hypothesis of heat death stems from Lord Kelvin's theory of heat and from the first two laws of thermodynamics extrapolated to a universal scale.

- A value $w > -1$ indicates a Big Crunch where the expansion of the universe eventually reverses and eventually re-collapses into a singularity (an infinite point of density) potentially followed by another big bang universe. An eternal cycle of big bangs if you will; nevertheless, the vast majority of evidence indicates that this theory is not correct. Instead, astronomical observations show that the expansion of the universe is accelerating, rather than being slowed down by gravity.

These are some of the things we need to understand to move forward with physics beyond the standard model and uncover the next laws of nature; although the way things go, when we finally figure out what dark matter-energy is made of, something even stranger will pop out. Nevertheless, careful study of the CBR is the best hope for the future.

Beyond the Standard Model

In 1900 Lord Kelvin stood in front of the British scientific society and pronounced: "*There is nothing else to be discovered in physics.*" He was very wrong; in the 1930s Max Planck came along and turned the world of physics upside down with quantum mechanics. Einstein, Pauli, Dirac, Born, Heisenberg, Schrödinger, and many more ran with it and created the golden age of physics. The trend continued in the 1950s and 1960s with the pioneers of quantum field theory, and the list goes on: inflation, dark matter, dark energy, string theory, quantum loop gravity, etc. Our universe is full of mysteries, and physicists are already looking at the next revolution. There are many instruments hard at work on this, but the most powerful is without a doubt the Large Hadron Collider (LHC).

LHC Is Back with a Vengeance

LHC is the Mecca for experimental physicists and the place to discover new laws of nature. In 2012 the LHC discovered the Higgs boson, and then it shut down for two years for upgrades but came back in 2015 with double the energy throughput it had when it discovered the Higgs. The goals of this new and improved LHC are twofold: (1) Have a better understanding of the Higgs and (2) discover new physics beyond the standard model. Physicists have great hopes and expectations about this. Here are some of them:

- Within the inner patterns of the standard model equation dwells a big mystery exposed when similarities were noticed between the equations of electromagnetism and the weak, strong nuclear forces. In other words, the three forces look similar, so we might wonder if in reality, the three are simply separate manifestations of the same super-force. Maybe there is only one force, and what we experience as three is in reality one thing looked from different perspectives.

- The calculations for the 12 matter fields, electron, neutrino, and quarks, are based on the same equation: the famous Dirac equation. So again, maybe there are no 12 matter fields in nature but a single field, and the reason we see 12 is because we are looking at the same thing from different perspectives.

Grand Unification and Supersymmetry: The Holy Grail of Physics

The ideas from the previous section go by the name of *grand unification*. That is the idea that electromagnetism and the weak and strong nuclear forces are the same super-force. However, there is another possibility: because both matter and forces are described by quantum fields, physicists have wondered if there is some way in which matter and forces are related to each other; the theory for that is called *supersymmetry*. Another bizarre theory out there proposes we just get rid of all the terms in the standard model equation and combine them all into a single term from which everything emerges: gravity, nuclear forces, particles, and the Higgs. – This is what *string theory* aims to find.

Tip   These ideas have driven theoretical physics for the last 30 years, and that is what the LHC aims to do. Right now there are no means to test string theory experimentally; however, grand unification and supersymmetry may be at the LHC grasp in the future. Altogether, the reason the LHC was built was to find the Higgs boson, but now with the job done, physicists think it is time to point the LHC's guns toward grand unification and supersymmetry.

Doom and Gloom in the Horizon

Since 2015 when a biffed LHC came online in search of grand unification and supersymmetry, the results have been disappointing: nothing has been observed. The physics community has been shell-shocked by this, and there is no consensus on what is going on; nonetheless, here are some of their responses:

- Have patience: We need a little more time; if nothing has been observed this year, perhaps something will show up in the future. In the end, if the LHC doesn't find something in the next few years, experts agree that it is highly unlikely that it will find anything at all.

- We are on the right track, but what we really need is a bigger machine to discover new physics. All in all, at a price tag of 4.4 billion USD for the LHC, and estimates around 10 billion for a bigger one to unravel grand unification. Not many governments in the world are willing to cover the price tag to explore these ideas.

- A third response is more speculative and not endorsed by many experts. Although physicists know that the standard model equation is correct, some believe that there are mysteries still hidden within it. Thus, perhaps the clues of a grand unification within this equation are just red hearings, and we need to get back to the drawing board to better understand it. This idea challenges the assumptions and paradigms of the last 30 years.

When We Are Wrong, We Start to Make Progress

Perhaps the lack of results by the LHC is what we need to get back to basics; sometimes we need to take a step backward to advance two steps forward. Some physicists feel energized by this lack of results because at the crossroads is where new ideas come along. For example, connections have been drawn between the standard model equation to other areas of science like condensed matter physics (the science of how materials work), quantum information science (the quest for quantum computer), and many more.

All in all, physicists are optimistic we can make progress, not in the way we wanted but progress nonetheless. The standard model equation may be king right now, but the hope is that we can come up with something better.

Exercises

Put your knowledge of basic quantum mechanics to the test with these easy exercises. If you get stuck, answers are provided in the appendix.

1. Fill in the blanks. The prominent physicist _____, said in _____: There is nothing else to be discovered in physics.

2. What German physicist is considered the father of quantum mechanics and what problem was he working on when he found a seminal discovery.

3. Write a small program to plot the experimental data for the ultraviolet catastrophe using the Rayleigh-Jeans law for spectral radiation of a black body as a function of its wavelength λ (lambda) and is temperature in Kelvin degrees (K) (Equation 1.1) vs. the adjusted Planck curves (Equation 1.2).

$$B_\lambda(T) = \frac{2cK_BT}{\lambda^4} \tag{1.1}$$

$$B_\lambda(\lambda,T) = \frac{2hc^2}{\lambda^5} \frac{1}{e^{\frac{hc}{\lambda K_B T}} - 1} \tag{1.2}$$

Use c = speed of light (299792458 m/s), K_B = Boltzmann constant = $1.38064852 \times 10^{-23}$ m^2 kg s^{-2} K^{-1} for temperatures of T = 3000, 4000 and 5000 K.

- Hint1: the x-axis of the plot is the wavelength λ. The y-axis is the spectral radiation B_λ.

- Hint2: The fastest way to do it is using the open source GNUPlot software; also save the plot as a JPEG image.

4. Give a brief description of the *photoelectric effect* experiment. What does the photoelectric effect indicate about the nature of light?

5. Who won the Nobel Prize for his new insight into the photoelectric effect? What was the old notion before this new insight? What was actually observed in the experiment?

6. True or false.

   a. The Copenhagen interpretation was the first and originated in the 1930s.

   b. The notion that energy travels not in continuous form but in discrete packets called quanta was postulated by Bernard Heisenberg.

   c. Niels Bohr is not the father of the *quantum jump*.

   d. The Rutherford model of the atom was the perfect and flawless description of the atomic structure at the beginning of the 19th century.

e. Albert Einstein said about the Bohr atom (quantum jump): *If I am to accept the quantum jump, then I am sorry I ever got into the field of atomic research.*

f. The father of the wave function ψ is Erwin Schrödinger.

g. The wave function now is considered the most powerful tool in physics in the last century.

h. German-Jewish physicist Max Born's description of the wave function sought to return to the continuous classical model of wave physics.

i. Schrödinger pushed the idea that all reality can be described entirely by waves and his wave function ψ.

j. Max Born came up with the *quantum cat* thought experiment to challenge Schrödinger.

7. Multiple choice: Which of the following are valid interpretations of quantum mechanics:

a. The star-gate interpretation

b. The many worlds interpretation

c. The Tijuana flats interpretation

d. The captain crunch interpretation

e. Quantum information

f. The Copenhagen interpretation

g. The quantum cat scratching the quantum dog interpretation

8. The uncertainty principle says what about the position and momentum of a particle?

9. True or false. The uncertainty principle gave rise to the notion of zero point energy.

10. What does zero point energy says about absolute zero?

11. In one sentence, define Pauli's exclusion principle.

12. Multiple choice: which of the following is true about Pauli's exclusion principle?

    a. The quantum cat used the exclusion principle to escape from the box.

    b. Pauli introduced a new two-valued quantum number that will later be known as the quantum spin.

    c. The exclusion principle was later generalized for all particles in the standard model.

    d. Bosons (photons, gluons, force carriers) do not obey the exclusion principle.

    e. The quantum dog does not obey the exclusion principle.

    f. Fermions (electrons, quarks, neutrinos) obey the exclusion principle.

13. The Dirac equation was an attempt of what?

14. In one sentence: What is the Dirac equation?

15. Which of the following is true about the Dirac equation?

    a. It predicts the existence of antimatter.

    b. It injects relativistic space time in the context of quantum mechanics.

    c. It follows Pauli's exclusion principle.

    d. It violates the uncertainty principle.

    e. It uses a two-component wave function with space-time coordinates (x,t).

    f. Dirac's equation has solutions with positive energies only.

16. Fill the blanks. In Dirac's Hole theory, holes in the _____ require _____ energies to create a particle-hole pair. The holes were later proven to be the _____, discovered experimentally by American physicist Carl Anderson in 1932.

17. In simple terms define particle entanglement.

18. What famous physicist called entanglement *Spooky action at a distance*? Why is it so spooky? Can we send text messages using entangled particles? If so, why?

19. Provide a brief description of Bell's theorem. Why is it important?

20. Multiple choice: Select the true statements about Bell's theorem:

    a. Defeated quantum mechanics.

    b. It was proven experimentally by French physicist Alain Aspect that quantum mechanics violates Bell's inequality.

    c. It proves that the properties of a particle are defined at the moment of observation.

    d. Einstein thought entanglement was a stroke of genius and came up with the EPR paradox to celebrate.

21. Fill in the blanks: the double-slit experiment shows the _____ duality of particles when a laser beam travels through a metal plate with two close slits with the resulting projection on a detector screen. If the measuring device is placed on the screen, the resulting interference pattern displays a _____. If the measuring device is placed on the slit, the interference pattern indicates a _____. This indicates that Quantum systems are fundamentally _____ (there are no exact answers) with the astounding implication that properties are defined at the moment of _____, instead of creation.

22. What are the differences between the Copenhagen and the many worlds interpretations of quantum mechanics?

23. In simple terms describe two additional interpretations of quantum mechanics besides the Copenhagen and many worlds interpretations.

24. Fill in the blanks. In quantum field theory (QFT), fields span the entire_____, the particles in the standard model are simply _____ in their respective field. Fields may _____ with one another.

25. Faraday is the father of what field? What was the original name he used for this field?

26. What are the four basic particle fields in QFT from which all particles in the standard model are defined? Which are the four force fields in nature?

27. What are the two partners of the electron in QFT? Why is it said that the electron field was reproduced twice?

28. What are the QFT partners of the neutrino, up quark, and down quark?

29. Why is the Higgs field called the god particle? What instrument was used in its discovery? At what energy level?

30. What is the Lagrangian of the standard model? Why is it important?

CHAPTER 2

Richard Feynman, Demigod of Physics, Father of the Quantum Computer

In the rankings of the greatest physicists of the last century, American physicist Richard Feynman (1918–1988) levels just a few notches below Albert Einstein. His name, legend, and legacy live in the halls of academia as the labors of a demigod. Feynman achieved so much in his 5+ decades in the field of physics: he was part of the Manhattan project that developed the first atomic weapon; his lectures and books in computation and quantum electrodynamics (QED) are the stuff of legend; his book *Six Easy Pieces* revolutionized the teaching of physics around the world. Feynman was one of the pioneers of quantum field theory (QFT), the modern view of quantum mechanics; his contributions were profound, earning him the Nobel Prize in 1965 for his work in quantum electrodynamics.

A little and obscure nugget about Feynman is that he was the first to propose the idea of a quantum computer. He realized that the tremendous complexity required to study phenomena at the atomic scale, ought to be performed by a machine designed with atomic principles in mind; that was the dawn of the quantum machine. This chapter is my take on the life and contributions of this remarkable individual.

© Vladimir Silva 2024
V. Silva, *Quantum Computing by Practice*, https://doi.org/10.1007/978-1-4842-9991-3_2

Mysteries of QFT: The Plague on Infinities

QFT has been fabulously successful in describing the smallest scales of reality, building on top of the Copenhagen interpretation of quantum mechanics to become the contemporary view of physics. However, this success has come at a cost, since even the most powerful equations in QFT such as the Dirac equation (see Chapter 1 for details) or Feynman's path integral become extremely difficult to solve even when used against the simplest particle interactions. Nonetheless, that hasn't stopped physicists in searching for clever ways to make them workable. Such efforts include

1. First, reduce the complexity by approximation into solvable form.

2. Next, deal with the irritating infinities that appear in these approximate equations.

3. Finally, group the whole thing into a pictorial system that most humans can understand; the grand result: Feynman diagrams.

To get an idea of how messy QFT can be, consider the simplest particle interaction: electron scattering (when two electrons bump into each other – see Figure 2-1).

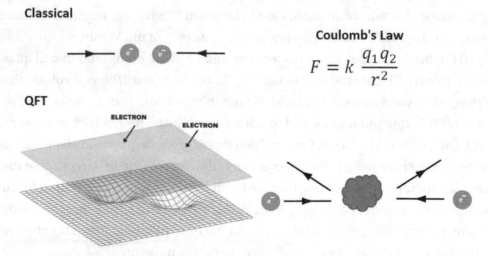

Classical

Coulomb's Law

$$F = k \, \frac{q_1 q_2}{r^2}$$

QFT

ELECTRON   ELECTRON

Figure 2-1. *(Top) Electron scattering according to classical electro dynamics, (bottom) the same process according to QFT*

In old fashioned classical electrodynamics, Coulomb's law quantifies the amount of force between two stationary, electrically charged particles. That is because each electron produces an electromagnetic field, and that field exerts a repulsive force in the other electron. The force is inverse to the square of the distance:

$$F = k\,\frac{q_1 q_2}{r^2}$$

where

- k is the Coulomb's constant, sometimes called electric force constant or electrostatic constant k = 8.99×10^9 Nm^{2}C^{-2} in the vacuum. Note that its value depends on the medium where the particles are immersed.

- q_1, q_2 are the charges of the electrons.

- r is the distance between them.

For mostly the simplest cases, Coulomb's law works well for this subatomic billiard shot, and it is easy to solve. However, in modern quantum field theory, specifically quantum electrodynamics (QED), the story is different. In QFT the electron is not a particle, but a vibration in the electron field (see Figure 2-1). Note that this field (and all quantum fields) permeates the entire universe, omnipresent whether or not there is an electron out there. Furthermore, the electron field may interact with other fields such as the electromagnetic (EM) field whose vibrations are responsible for producing the photon or what we perceive as light. This is how QED describes electron scattering.

Electron Scattering According to QED

A vibration in the electron field produces what we identify as a particle; this field in turn is connected to the EM field (vibrations in one may produce vibrations in the other). When the electron excites the EM field, it produces a photon which in turn delivers a bit of the momentum from the first electron to the second. In QFT, the exchanged photon is called a virtual particle, and its existence is clouded in mystery as a result of the quantum event, suffice to say that this virtual photon exists long enough only to communicate the force. There are other types of virtual particles whose existence is equally ambiguous but more on that in the next section on Feynman diagrams.

This is the perfect time to get our feet wet with our first Feynman diagram, one for electron scattering according to QED. We'll go under the hood further down.

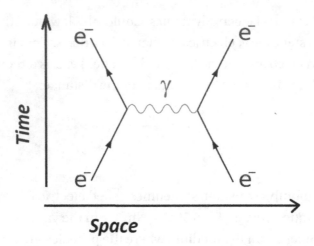

Figure 2-2. *Feynman diagram for the most probable case of electron scattering*

In Figure 2-2, we see two electrons traveling toward each other in a coordinate system where the Y-axis represents time and the X-axis represents space. Note that the actual distances in the space axis are not relevant:

- The electrons exchange a virtual photon (denoted by the Greek letter Gamma γ), that is, the 3 point vertex with the squiggly line.

- The electrons move away at the end.

But Feynman diagrams aren't just drawings of the interaction, as we'll see in the next section, but equations in disguise, with each part of the diagram representing a chunk of the maths. Where things get really bizarre is at the vertex with the squiggle (the virtual photon exchange), so much saw that QFT states that:

Tip   There are infinite ways electron scattering can occur; furthermore, all intermediate events that lead to the same result do happen. Thus, to perfectly calculate the scattering of two electrons, we need to add up all the ways this could happen.

With infinite possibilities behind this simple process, a complete quantum field theory solution is not possible. Enter perturbation theory, an absolutely indispensable tool to solve QFT problems.

Perturbation Theory: If You Can't Do Something Perfectly, Maybe Near Enough Is Good Enough

Perturbation theory is the art of solving unsolvable equations: If you have an impossible equation, just find a similar equation that is solvable and make small changes to it (perturb it) so the result is similar. It will never be exact but it will get you close enough.

In the case of electron scattering, it turns out that the most likely interaction is the exchange of a single virtual photon. All other interactions contribute less to the probability of the event.

Tip   Richard Feynman said that the more complicated the interaction, the less it contributes to the final probability amplitude, and here is where his diagrams come in handy; as a matter of fact, it turns out that the probability of a particular interaction depends on the number of connections (vertices); so much so that each additional vertex in an interaction reduces the final probability by a factor of around 100.

Thus, the most probable interaction for electron scattering is the simple case in Figure 2-2 with two vertices and a single virtual photon exchange; a three vertex exchange would contribute about 1% of the probability of the two vertex interaction, a four vertex interaction would contribute about 1% of 1%, and so on. Perturbation theory and Feynman diagrams help identify the important additions to the final equation, and which interactions we can ignore. They make the complex calculation possible, but we are not done yet.

The low probability of intermediate states opens a new enigma which is especially true for the *loop interactions*. Let's take a closer look.

Tackling Those Pesky Infinities with Renormalization

Loop interactions are weird intermediate states that wreak havoc in Feynman diagrams. Two quintessential examples of loop interactions are (see Figure 2-3)

1. A photon momentarily becomes an electron, positron pair, and reverts to a photon again.

2. An electron emits and then reabsorbs the same photon.

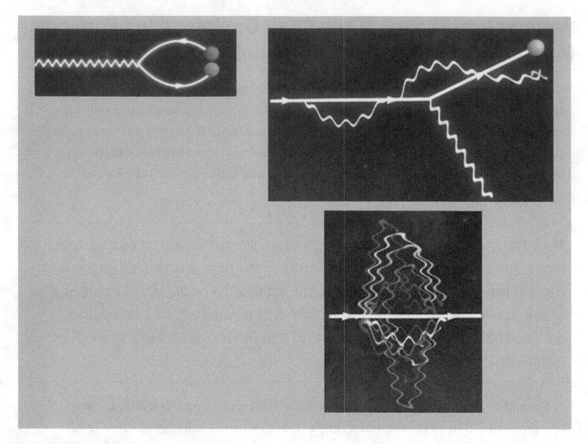

Figure 2-3. *Two examples of loop interactions: on the left a photon creates an*
electron, positron pair, then reverts to a photon; on the top right, an electron emits
then reabsorbs the same photon; on the bottom right, the effects self-energy loops
due to virtual photon interactions

Tip   For the second case, QED states that the electron causes a constant
disturbance in the EM field, due to the fact that electrons are constantly interacting
with virtual photons; this slows down the electron and increases its effective mass.
This effect is known as *self-energy*.

When trying to calculate the self-energy correction of the electron mass using QED,
the result is infinite extra mass. Things don't look good for QED thus far, but why is it
infinite? QED says that to calculate the mass correction from these self-energy loops, we
need to add all possible photon energies, but those energies can be arbitrarily large, thus
sending the corrected mass to infinity and beyond.

Tip   Physicists have speculated that in reality, self-energy loops must have a limit, but we don't know what that limit is; yet another mystery in our crazy universe, although some say that the answer lies in quantum theory of gravity which we don't yet have.

All in all, just like with perturbation theory, physicists have found a cunning solution to keep in check the infinities in self-energy loops: renormalization.

Renormalization: Electrons Do Not Have Infinite Mass

Renormalization is the second tool in QFT's arsenal against the infinity horde. We know that electrons don't have infinite mass as its value has been calculated experimentally, and here is where the renormalization trick comes into play (see Figure 2-4).

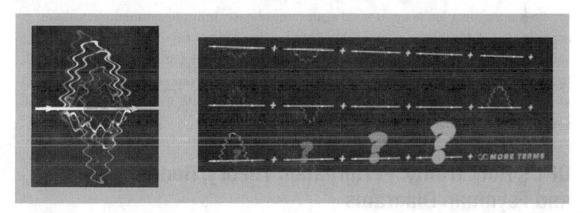

Figure 2-4. *Self-energy loops slow down the momentum of the electron and send its effective mass to infinity*

Renormalization says: don't start with the corrected (or fundamental) mass of the electron – it is unmeasurable – use the *experimentally measured mass* instead. In other words, don't use its theoretically calculated value but its experimental (finite) number, and then solve the equations from there. This clever trick can be used to eliminate many of the infinities that arise in QFT! Here are two specific scenarios where renormalization is an essential tool:

1. Calculation of the corrected mass of the electron due to self-energy loops

2. Infinite shielding of electric charge due to a virtual particle, antiparticle pairs popping in and out of existence

---

Note   Renormalization works because experimental measurements of the mass of the electron include some of its self-energy loops. Consequently, this measurement is not that far off its fundamental or bare mass. With that in mind, renormalization can only make predictions of a particle's mass relative to its lab measurements.

So renormalization is a god-send against those nasty infinities; nonetheless, there is a price to pay for this rule: For each infinity you want to get rid of, you must measure some property in the lab, which means the theory can't predict that particular property from scratch; it can only predict other properties relative to your lab measurement. All in all, renormalization saved QFT from the plague of infinities. We finish the trinity with the super slick Feynman diagrams!

QFT's Holy Trinity: Perturbation Theory, Renormalization, and Feynman Diagrams

In QFT, all subatomic particles are expressed as vibrations in quantum fields. But even the most elegant formulations become impossibly complicated when used in anything but the simplest particle interactions. Richard Feynman gave us a simple set of pictorial rules to manage this complexity. His diagrams have become one of the de facto tools in contemporary physics. They successfully describe things like

- Particle scattering

- Self-energy interactions

- Matter, antimatter creation and annihilation

- All sorts of decay processes

Feynman diagrams are an incredibly powerful tool to predict the behavior of the subatomic world and are a central part of the standard model of particle physics.

Feynman Diagrams: Formulas in Disguise

Feynman diagrams revolutionized theoretical physics by providing a pictorial representation of the infinite possibilities of particle interactions. Feynman diagrams are awesome because

- Their beauty and simplicity provide striking insights into the nature of reality.

- Behind their pictorial look and feel hides a mathematical formula about particle interactions. This simplifies things enormously to the point of making the tedious and complicated mathematics of particle physics enjoyable and intriguing.

- They are a tool to crack an impossible equation: the infinite possibilities hidden between two measured states of a particle.

Feynman Approach to Quantum Mechanics: The Path Integral

Feynman's remarkable work on the path integral earned him a Nobel Prize in physics in 1965. His brainchild, the path integral, states that to properly calculate the probability of a particle traveling between two points A and B, we need to add the contributions from all conceivable paths between those points (left side of Figure 2-5) including the impossible paths (right side of Figure 2-5).

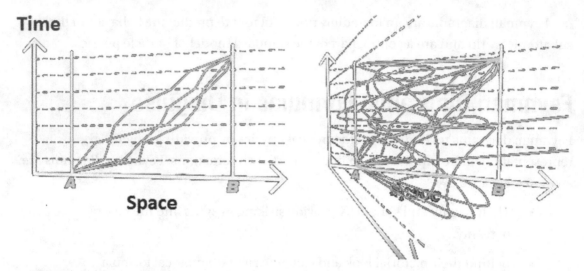

Figure 2-5. *In the path integral every conceivable happening from a measured initial state to a measured final state does happen in the maths*

Unraveling the Impossible: Feynman Diagrams to the Rescue

The path integral tells us that there are infinite possible intermediate states in a quantum system evolving between two states. Furthermore, to calculate the probability of this evolution, we need to add all conceivable intermediate states. This is simply not possible; however, here is where Feynman diagrams shine at their brightest: they allow physicists to figure out which possibilities need to be considered to get an acceptable answer.

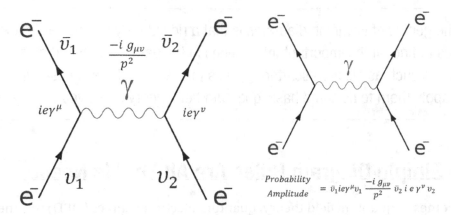

Figure 2-6. *Feynman diagram showing electron scattering, or the interaction of two electrons (e⁻) resulting in the emission of a photon (γ)*

Each diagram represents a family of particle interactions (left side of Figure 2-6) and gives the equation to calculate the contribution of that interaction to the total probability (right side of Figure 2-6). In this particular case, we use a diagram for electron scattering where

- v_1 v_2 $\bar{v}_1$ $\bar{v}_2$ represent the incoming and outgoing charges of electrons 1 and 2, respectively.

- $ie\gamma^\mu$ $ie\gamma^\nu$ represent the absorption and emission of the photon (these are connecting points or vertices).

- The squiggle $\dfrac{-i\,g_{\mu\nu}}{p^2}$ is the quantized field excitation of the photon.

The equation coming out of this diagram represents all the ways two electrons can bounce off each other, involving a single photon, and from this equation, it is possible to calculate the effect of the exchange. Unfortunately, electron scattering at the quantum level is a lot more complicated than this.

When we observe electron scattering, all we see are two electrons bouncing off each other; however, the quantum event at the center of the scattering remains a mystery. Quantum mechanics says there are infinite ways that scattering can occur, and the weirdest part is that they all happen at the same time!

Tip   The genius of Feynman diagrams is that a ridiculously simple set of rules allows us to find all the important interactions in the path integral and other QED interactions such as electron scattering. Let's take a look at these easy rules and how to apply them to do some basic quantum field theory ourselves.

A Few Simple Diagram Rules Are All That Is Needed

When it comes to quantum field theory, quantum electrodynamics (QED) was the first, and it is currently the most powerful of the field theories out there. QED studies the interactions of the electron field with the electromagnetic field, that is, interactions among the electron, its antimatter partner the positron, and the photon.

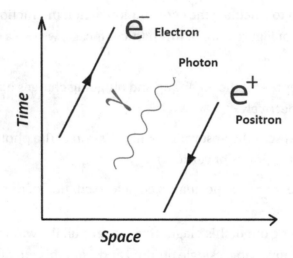

Figure 2-7. *Building blocks of a Feynman diagram*

In a Feynman diagram, the electron (e⁻) is depicted as an arrow pointing forward in time, while the positron (e⁺) is an arrow pointing backward in time. Note that depicting antimatter as backward in time looks strange, but it is by design and provides a very powerful mechanism as we'll soon see. Finally, the photon (γ) is shown a wavy line without any time direction. Stick these three elements in the space time coordinate system, and we have a basic Feynman diagram, a useless one nonetheless.

To start realizing the power and simplicity of this approach, the electric and electromagnetic fields need to interact. Things start to get interesting when particles interact; this interaction is depicted as a vertex in the diagram. A vertex is the point where the lines representing the different particles come together.

Tip   It turns out that, in QED, only one vertex is possible: one with an arrow pointing in, one with an arrow pointing out, and a single photon connection (see Figure 2-7). This vertex alone represents six different interactions and can be used to construct infinite Feynman diagrams.

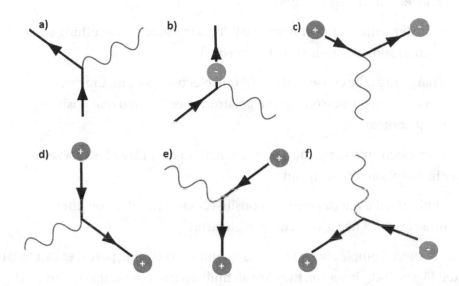

Figure 2-8. *Possibilities from a single incoming, outgoing electron-photon vertex in a Feynman diagram*

In Figure 2-8, we see the possible interactions from the basic vertex diagram (a) with time increasing upward:

a.   This diagram represents an initial electron that emits a photon; afterward both particles move in opposite directions; now if we rotate the vertex so the electron and photon come from below (b), then

b.   This diagram represents an electron absorbing an incoming photon, the photon vanishes, and its momentum is completely transferred to the electron. Rotate again then,

c.   The picture is of a photon coming in and giving up its energy to produce an electron-positron pair (a process that is called pair production). Rotate again then,

d.   Here we have a positron absorbing a photon,

e.   A positron emitting a photon, and finally,

f.   An electron and a positron annihilate each other to produce a photon. These are all the ways the electron and electromagnetic fields can interact.

Every single QED interaction is built from Figure 2-8. Note that Feynman diagrams must obey conservation laws including

- Particles cannot appear nor vanish from the void (if something goes in, then something else must come out).

- Charge must be conserved too (if one electron goes in, the one electron must come out; if one positron goes in, then one positron must come out).

- If an electron and positron both go in, then they cancel with a zero charge photon coming out.

- Similarly, if a photon creates a positively charged positron, then it must create a negatively charged electron.

There are more complex ways in which incoming-outgoing particles can balance charge (see Figure 2-9); however, they are all built up from the single vertex (right side a Figure 2-5).

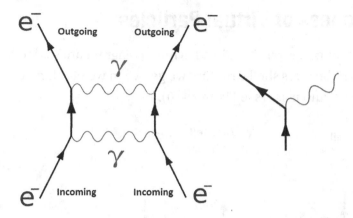

Figure 2-9. *Endless diagrams can be constructed from a single three path vertex in QED*

A final and perhaps one of the most important rules is that the overall interaction described by a set of diagrams is defined by the incoming particles (particles going in) and the outgoing particles (particles going out). These are the particles to be measured (e.g., for energy or momentum), and they must obey the energy-momentum relation:

$$E^2 = \bar{p}^2 c^2 + m^2 c^4$$

where E is the object's total energy, m is the rest mass, and p is the momentum.

Tip   The *energy-momentum* relation is a relativistic equation closely related to Einstein's famous *mass-energy* equation E = mc², and it was first established by

Paul Dirac in 1928 as $E = \sqrt{c^2 p^2 + \left(m c^2\right)^2} + V$ where V is the amount of potential

energy. Dirac used the energy-momentum relation to describe the behavior of the electron in relativistic terms and in doing so took the first crack at unifying special relativity with quantum mechanics; a feat that earned him world fame, a Nobel Prize, and a place in history.

The Strangeness of Virtual Particles

The incoming and outgoing particles in a Feynman diagram are said to be *on-shell* because they sit on the mass shell structure we get when we plot Einstein's equation of energy, momentum, and mass (see Figure 2-10).

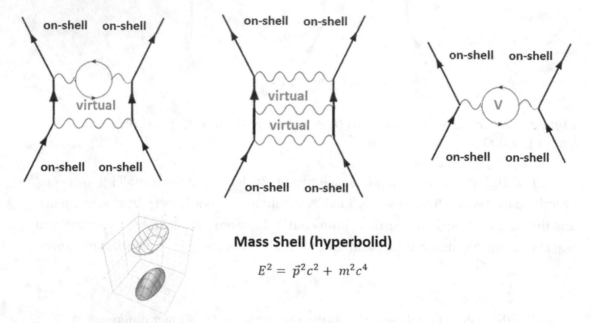

Mass Shell (hyperbolid)

$$E^2 = \vec{p}^2 c^2 + m^2 c^4$$

Figure 2-10. *Set of Feynman diagrams showing on-shell particles along with virtual particles (top), plus a hyperboloid surface (shell) of the equation for energy, momentum, and mass*

On the other hand, everything that goes on between the incoming and outgoing tracks of the diagram is said to be virtual, and the possibilities are endless. That is, each possible diagram that results in the same set of incoming-outgoing particles is a valid path for the possibility space of that interaction (see top of Figure 2-10).

Tip   Virtual particles exist between vertices within the diagram but don't enter or leave. They are also by definition unmeasurable.

Other peculiar characteristics of virtual particles include

- They do not obey the mass-energy-momentum equivalence; thus, they are called *off-shell*.

- They are not limited by the speed of light or the direction of time! This leads to the endless possibilities mentioned before.

The Power of Feynman Diagrams to Simplify QFT-QED Calculations

We saw from the previous section that the set of diagrams for the simplest electron scattering interaction can be endless. As long as the end result is the same, all possibilities must be considered. The genius of Feynman diagrams is that they are an incredibly powerful tool in simplifying QED interactions. This is because the interpretation of the interactions is irrelevant; all we care about is the topology of the diagram, which is how the vertices are connected to each other.

Tip   Feynman diagrams simplify QFT calculations by reducing the number of contributing interactions that need to be solved. For example, for two electrons exchanging a single photon (see top of Figure 2-7), it doesn't matter if we draw the photon going from the first vertex to the second or the second to the first, even though both seem like two different interactions. Feynman thought of this difference simply as the photon traveling forward in time in the first case and backward in time in the other. The maths covering the transfer is the same in both cases. A weird yet powerful characteristic of Feynman diagrams.

QED is a really tough nut to crack and here is why:

- Each diagram represents an infinite number of specific interactions. For example, each of the particle paths is actually infinite, as well as infinite possibilities for particle momenta.

- All possible paths must be considered, even the impossible ones, such as faster than light paths, something that is a big no-no in Einstein's relativity.

- For any virtual particle (off-shell, neither incoming nor outgoing), any energy, speed, and even direction of time must be considered, and by considered we mean mathematically solved.

This last point that any *direction of time* must be considered in QED interactions is where Feynman diagrams shine at their brightest. To illustrate this point, let's look at another quintessential electron interaction, one called Compton scattering (see Figure 2-7).

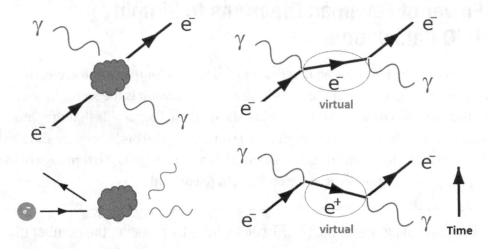

Figure 2-11. *Compton scattering where an electron and a photon bump into each other*

In Figure 2-11 an electron and a photon bounce off each other. On the left side we see the on-shell electron and photon; we can measure properties on these two such as momentum or energy, but whatever goes on off-shell (within the cloud) cannot be measured and that falls in the realm of the bizarre. QED says there are infinite possibilities for whatever happens within, and they all must be solved, even the impossible ones such as paths backward in time. It looks like an impossible task, but not so when Feynman diagrams are used. Feynman diagrams can easily represent paths backward in time using antimatter. Consider the right side of Figure 2-11 where we have diagrams for two different paths of Compton scattering:

- One way this may occur is if the electron emits a new photon and later absorbs the incoming photon (top right of Figure 2-11). In the intermediate stage between vertices, the electron is a virtual particle which means we include all possible paths it may take as long as the result is the same (including paths backward in time).

- According to Feynman an electron traveling backward in time looks like a positron; thus, an impossible path can be easily represented as follows (bottom right of Figure 2-11): The same particles go in and out (as in the previous option); however, the interaction looks a little different. Instead of an electron emitting a new photon, we have an incoming photon emitting a positron-electron pair. The new electron becomes our outgoing electron, and the positron annihilates the incoming electron to produce the outgoing photon.

These may seem like two different interactions; however, in the world of Feynman diagrams, the maths is the same; whatever happens off-shell is irrelevant as long as the result is the same. We don't care about the interpretation of an interaction in the diagram, only about its topology (how the vertices are connected). All in all, by using Feynman diagrams, we end up vastly reducing the number of contributing interactions that need to be solved in QED, simplifying the complex maths required in QFT calculations.

In the next section, we look at how fabulously ingenious the idea of describing antimatter as time reversed matter is, and how it saved physics from its biggest threat yet: symmetry violations.

Antimatter As Time Reverse Matter and the Mirror Universe

Comic book lore is full of parallel dimensions, multiple universes, and that sort; in that context, there is a seminal story line in DC comics called *Crisis on Infinite Earths*. In that story there is an all-powerful being, the *Monitor*, who can bend the laws of time and space to his will, shaping reality as he wishes. The Monitor wages war against an evil, equally powerful but symmetrically opposite being: *The Anti-Monitor*. This evil entity unleashes a wave of antimatter to destroy all universes and all Earths within them. In the end, the antimatter wave destroys the multiverse; however, the Monitor gathers the greatest heroes of the multiverse for a final battle against the Anti-Monitor. In that climax, the Anti-Monitor is defeated, and our heroes can reboot the multiverse releasing an event that looks a lot like a quantum fluctuation (or a big bang). Finally, the multiverse is saved, although not exactly as before, and life goes on.

Not a far-fetched story by any means, as there are interpretations of quantum mechanics that include infinite realities (Earths), multi-verses, quantum fluctuations, and other bizarre stuff. All in all, in our reality there is no such thing as an Anti-Monitor or an antimatter wave annihilating everything in its path; nevertheless, that didn't stop our universe's physicists on imagining a mirror universe where the laws of physics work the same but in opposite direction; Feynman called this universe parity symmetric.

The Foundations of Quantum Theory Rest on Symmetries

Look at yourself in the mirror; could that be a symmetric version of you? Now imagine your reflection is made of matter of the opposite charge; not only that but imagine time flowing in the opposite direction within the mirror, if that reflection could be yanked into reality then, your alter ego would be a perfectly symmetric you. But let's start at the beginning, what is the first thing that comes to your mind when you hear the word symmetry?

In the classical definition of symmetry, we may think of reflective or mirror symmetry: try this, draw a square or a circular face in paper, then draw a vertical axis in the center, fold the pieces, and if they are identical, then you have a symmetric shape. That is known as geometrical or reflective symmetry (see Figure 2-12).

Tip   In physics, symmetry results when the laws of physics are unchanged under a transformation of coordinates; our universe is symmetric to coordinate shifts in space, time, and when it comes to the atomic scale, under the abstract phase of the wave function in quantum mechanics.

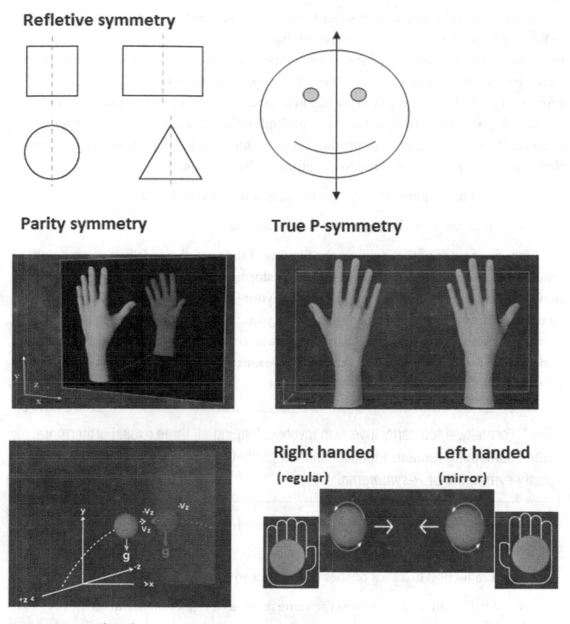

Figure 2-12. *(Top) Examples of continuous or reflective symmetries; (middle) discrete parity symmetry given by translation in the spatial axes (XYZ); (bottom) handiness in a regular versus a mirror reflection*

There are ways, however, under which our universe is not symmetric; as a matter of fact, the laws of physics are not the same for a mirror universe, but to understand this we need to understand the meaning of a *parity transformation*.

In the realm of physics, the geometric symmetries at the top of Figure 2-12 are said to be *continuous* because there is no such thing as a partial reflection; you either reflect the whole thing or you don't (think of symmetry as a motion or change in position over time). Parity, on the other hand, is a *discrete symmetry*. A discrete symmetry describes non-continuous (disconnected) changes in a system. For example, a square has discrete rotational symmetry, as only rotations by multiples of right angles (90 degrees) will preserve the square's original appearance (e.g., rotate the square 45 degrees to become a rhombus). Other types of discrete symmetries in physics include

- Charge conjugation: Flipping the sign of the electric charge

- Time reversal: Having a clock tick backward

A parity transformation involves the flipping of the spatial axis. When you put your hand in front of a mirror, you have a parity transformation in the z-axis; your reflected hand has turned back to front, and it looks like your left hand becomes your right and vice versa. The weird thing is that it seems that your hand flipped in the x-axis, not the z-axis. As a matter of fact, a true x-axis transformation results when left becomes right and forward facing stays the same, that is, a true x-axis transformation equals a z-axis reflection plus a rotation (see Figure 2-12).

Tip   Formally, a full parity inversion involves flipping all three axes. Furthermore, when something remains identical under a parity transformation, it is said to be parity symmetric or *P-symmetric*.

Also, a reflection in any single axis can be reproduced by a reflection on a different axis plus a rotation:

- A reflection in 2 axes pushes things back to the way they were.

- A reflection in 3 axes gives the same result as a reflection in one axis.

Now, in the case of our reflected universe, if we throw a ball to the mirror, then it looks like it is coming toward us. Here we realize that even though some properties flip signs in the mirror, others remain the same (see Figure 2-12). Thus

Properties that flipped called *odd-functions* are

- Spatial axes: X, Y, Z

Properties that do not change in the mirror (*even-functions*) are

- Angular momentum

- Energy of the ball, its mass, gravity

- Time in the clock

Tip   The laws of physics should work the same in the mirror world even if the spatial axes get flipped. For example, Newton's laws of motions can be used to calculate the ball's velocity and trajectory, but that is true only if the ball remains unchanged in the reflection.

Now imagine that we add spin to the motion of the ball (see bottom right of Figure 2-12); notice that this spin or angular momentum remains unchanged in the reflection, but yet the momentum vector or direction of motion is flipped. This is similar to the switching of left-right hands in a mirror reflection and introduces the critical concept of *handiness*.

Note   Picture grabbing the ball with the fingers pointing in the direction of the spin, and the thumb pointing in the direction of motion. If we use the right hand to do it, then we have right-handed spin relative to the motion vector; if we use the left hand, then we have left-handed spin. *In physics, the handiness of a spinning moving object is called chirality.*

Chirality in quantum mechanics has a specific meaning nonetheless: *quantum chirality is related to quantum spin where motion doesn't factor in.* Instead, it is considered to be a fundamentally internal left- or right-handed asymmetry. Most particles have both left- and right-handed versions, and it was believed at the beginning that both behaved equally. This was the consensus given by early experiments involving gravity, electromagnetism, and the strong nuclear force. At the time, the same was thought to be true for the yet untested weak nuclear force. As we'll soon find out, this turned out to be untrue.

Richard Feynman, in his famous lectures on physics, talked about such a mirror universe and described what it would take for it to be parity symmetric. He imagined a clock in a mirror reflection where numbers are backward, components are flipped left to right, and it ticks counter clockwise. Finally he invited us to build that clock in our reality:

- By constructing every piece of the clock as though reflected: Numbers painted backward.

- Every right-handed screw or coil is replaced by a left-handed version and vice versa.

Finally he asked: Will this mirror clock tick the same exact way counter clockwise? Our intuition tells us that it should, as we replaced all mechanical components with their mirror opposite; unfortunately this is wrong; the mirror clock will still tick clockwise.

Tip   A parity transformation is obtained by substitution of the properties of an object with their mirror opposite: Positive charge becomes negative, left becomes right, up becomes down, and so forth. The laws of physics are not symmetric to this type of parity transformation.

This fact was proved experimentally in 1956 by the brilliant Chinese-American experimental physicist Chien-Shiung Wu (1912–1997) using the radioactive isotope of Cobalt (Cobalt-60) atom.

Wu started with the simple argument, that if parity is conserved, then there should not be an experiment to determine if we live in the regular or the mirror universe; in other words, it makes no difference which universe we find ourselves in as long as parity is conserved. Thus, she came up with an experiment to test if parity is conserved in our universe. It is slick and works as follows:

The Cobalt-60 radioactive isotope decays into nickel by the effect of the weak nuclear force, emitting an electron, some gamma ray photons, and neutrinos in the process. Its nucleus also happens to have an unusually strong spin, so Wu and colleagues used this fact to align the axis of the spin of a set of atoms using a magnetic field and watched them decay (top of Figure 2-13). They found that the electrons produced in the decay emerged in the opposite direction on the spin. This was the smoking gun that proved that parity symmetry is violated in our universe. Let's see why.

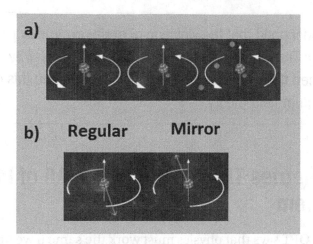

Figure 2-13. *(Top) Atoms of Cobalt-60 with spin aligned upward using a magnetic field ejecting electrons in the opposite direction. (Bottom) Momentum of the ejected electron for Cobalt-60 in a regular vs. mirror reflection as proof of parity violation*

Tip   The fact that the spin of the nucleus is correlated to the momentum of the ejected electron is the clue that convinced Wu that parity transformation is violated. This is because the spin of the nucleus (represented by the angular momentum vector along the axis — or the arrow pointing up off the nucleus) in a reflected universe does not change (spins in the same direction, pointing up in both cases); however, the ejected electron momentum points in the opposite direction for the mirror reflection (bottom of Figure 2-13). It turns out that for parity to be conserved, *there must not be any correlation* between the direction of the spin of the nucleus and the direction of the ejected electron. This way nothing would change in the reflection, yet that is not what the experiment showed.

Feynman used Wu and colleagues' experiment to propose a clock governed by the decay of Cobalt-60 using the spin of the aligned atoms and adding a detector at the bottom which ticks every time an electron is ejected downward. He then showed that in a reflected clock where the Cobalt atoms are replaced by the parity inverted counterparts, the decaying electrons travel upward instead, away from the detector. He concluded that a perfectly constructed Cobalt-60 mirror clock would never tick at all.

Tip Feynman understood that the violation of parity symmetry poses a threat to a higher symmetry: The so-called CPT symmetry or *charge-parity-time* symmetry (that is the combined flipping of charge, parity, and time); and this one lies at the foundation of quantum field theory.

Broken Symmetries Threaten to Break All of Physics Along with Them

The absolute rule in QFT says that physics must work the same if we flip these three properties (charge-parity-time), else all hell breaks loose and QFT along with it. With parity already violated, Feynman sought to save QFT by showing that charge-parity or CP symmetry could be saved. He proposed a modified Cobalt-60 clock made of antimatter where

- Electrons become positrons.

- Quarks become antiquarks sending protons and neutrons to their anti-counter parts.

- Cobalt-60 becomes anti-cobalt-60.

Tip Charge conjugation (flipping) is the C-part in a CP transformation. That is what switching to antimatter means.

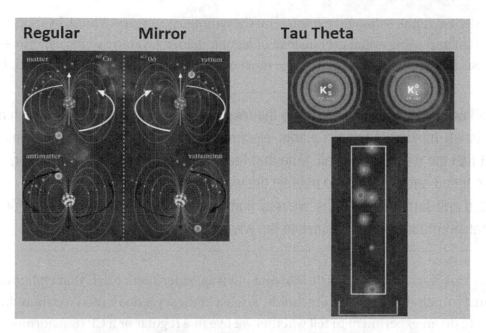

Figure 2-14. *(Left) Feynman's Cobalt-60 matter, antimatter clock showing how CP symmetry holds in a mirror universe. (Right) Tau Theta experiment by Cronin and Fitch to test for CP violations*

Can CP Symmetry Be Saved by a Clock?

In Figure 2-14 we see Feynman's Cobalt-60 antimatter clock in action; on the left side the regular universe and on the right its mirrored reflection:

- On the top left, the regular matter Cobalt-60 atom with its spin axis aligned to the top by a magnetic field, and at the bottom, its antimatter counterpart. Because the antimatter atom has negatively charged nuclei, its nuclear magnetic field will point in the opposite direction relative to regular matter. Thus when the magnetic field is applied, the antimatter Cobalt-60 will align to the bottom as shown.

- In the mirror reflected antimatter clock, both the direction of the decaying electrons, and the direction of the spin are flipped; once due to the mirror reflection, and one due to the switch to antimatter.

This leaves the electrons in the mirror, traveling in the original direction – down, and the clock ticking as usual. All in all, even if our universe is not parity symmetric, *maybe it is symmetric under a charge-parity (CP) transformation.*

Under Feynman's thought experiment, CP symmetry appears to hold, by flipping left to right and using antimatter instead of regular matter. Over the years, experimental physicists have shown this to be true for most of the particles in the standard model.

Tip   Parity symmetry violations are the result of the weak atomic force which only affects left-handed fermions (quarks, electrons, neutrinos). Right-handed fermions don't feel the weak force at all. Note that the opposite is true about antimatter: Left-handed anti-fermions do not feel the weak force, while right-handed anti-fermions do. Thus, we may find ourselves in the same situation for a CP transformation when it comes to the weak force.

All in all, the key to this puzzle is to find out if an experiment exists that violates a CP transformation. If CP symmetry holds, at least in theory it does, then we shouldn't be able to do an experiment to tell whether we live in a regular or a CP transformed universe. All physics must work the same in both.

Strike 2: CP Symmetry Is Violated, Three Strikes, and Physics Is Ruined

In the cosmic game of broken symmetries, strike 1 was dealt when experimentalist Chien-Shiung Wu showed that parity transformations are violated in the Cobalt-60 radioactive isotope. Now we find ourselves at the bottom of the 9th, all bases covered, and CP symmetry in the spot. With the roar of the crowd, we seek to find out if an experiment exists that violates CP transformations. Physicists started to worry when such a thing was found around the so-called *Tau Theta Problem*.

Tau and Theta are two particles that were thought to be the same; they have the same mass, electric charge, and spin, but decayed into different products: Tau decays into three Pions while Theta decays into two Pions, that being the only difference. Furthermore Tau's decay products have odd parity while Theta's have even parity (this implied a violation of parity conservation, thus giving rise to the so-called Tau Theta problem). In 1956 physicists Tsu-Dao Lee and Chen Ning Yang proposed a solution: They claimed that Tau and Theta are in reality the same particle, known today as Kaon (K^+), and that the weak nuclear force responsible for its decay, does not preserve parity. This solution earned them a Nobel Prize in 1957.

Later on, in 1964 American nuclear physicists James Cronin (1931–2016) and Val Fitch (1923–2015) looked at the outcomes of the decay of a weird form of Kaon called *Neutral Kaon* looking for hints of parity violations.

Tip   Neutral Kaons are a weird quantum mix of its own particle, anti-particle. This mix exists in two states: K_S^o (K-S) is short lived and doesn't change under a combined charge-parity transformation (CP: even). The second type K_L^o (K-L) is long lived and has an odd CP state (it changes under a CP transformation). *If CP symmetry is to be conserved, K_S and K_L should never transform into each other because they have different CP states.*

Cronin and Fitch found out that Neutral Kaons display a bizarre behavior in this universe: they sent a bunch of both types of Neutral Kaon down a tube into a detector in the far end (right side of Figure 2-14). The short lived K_S particles should never have completed the journey given their short lives, and yet, a significant number of decay products from K_S particles were found at the detector.

Note   The only explanation for Cronin and Fitch result is that the long lived K_L particles transformed into K_S particles violating CP symmetry. It looks like Feynman's mirror reflected antimatter Cobalt 60 clock is a bust after all!

Time Symmetry Conservation: The Hallowed CPT Looks in Danger

The violation of CP symmetry has dire consequences for the laws of physics; the only remaining hope in the C-P-T trinity is *time reversal symmetry*. All the results so far suggest that time symmetry is broken as well, but for that, we need to get back to the past, to the beginnings of quantum field theory.

We find ourselves in the 1950s, at the dawn of QFT; at the time it was thought that this new promising theory required certain symmetry. As time went on, and the theory developed, it became clear the demand for symmetry under the combined action of **c**harge conjugation, **p**arity inversion, and **t**ime reversal:

- The axiomatic foundations of QFT state that an antimatter, mirror reflected, and time reversed version of our universe should have the same laws of physics (this is known as the CPT theorem which postulates that QFT must be CPT invariant).

- CPT invariance was the definite view of physicists throughout the 1950s: The laws of physics must work the same under a flip of charge, parity, and direction of time, but then experiments popped out that showed the violation first of **p**arity, then **CP** symmetry, so we are left holding to our last hope: T-symmetry.

Tip   If CPT symmetry holds (to save the world of physics) but CP symmetry is violated, then T-symmetry must also be violated. This is because the time reversal transformation needs to bring us from a broken CP mirrored universe into a stable CPT universe.

The problem with the statement above is that if we go the other way around, with a T-transformation from the stable CPT to the broken CP mirror universe, then the time reversal transformation changes the way the universe behaves. That is really bad, as physics is supposed to work the same whether we go forward or backward in time. So T-symmetry goes out the window; of course this is all theoretical; nothing has been set in stone.

Note   Time reversal symmetry is required by quantum mechanics in order to conserve quantum information. Quantum mechanics says that information cannot be destroyed.

So what is it? Is T-symmetry violated or not? The answer depends on what we mean by time reversal:

1. The most obvious interpretation of time reversal is when we flip the so-called *arrow of time*, having the universe travel backward in time. Presumably under this T-transformation quantum information is preserved.

2. A second interpretation for time reversal is shown in the right side of Figure 2-15. This interpretation is thought of as flipping the direction of the evolution of a physical system: For example, an explosion becomes an implosion, or particle decay becomes particle creation. In other words, *don't rewind time nor convert matter to antimatter, but reverse all momentum and spin.*

Tip   Option one is not what physicists meant by T in CPT. In reality, the correct T comes from option two where all momentum and spin are reversed, that is, taking all particles and pointing them back to where they came from. If this T-symmetry holds then, after reversing all particle motion, these particles should retrace their steps, reactions, and histories. Furthermore, as this motion-spin reversed universe evolves forward in time, it should end up in its starting arrangement. On the other hand, if this T-symmetry is broken, then the future will not perfectly mirror the past.

Physicists argue that one prediction of this T-symmetry is that all processes should take the same amount of time going forward as going backward. If that is true, then an experiment could be created to put this to the test by timing the quantum transition between one particle type and another; it should be the same in either direction. As a matter of fact, in 2012, physicists from the Babar Collaboration at the Stanford linear accelerator tested the speed at which B-mesons transition between two types. If this T-symmetry holds, then the speed should be the same in either direction; their result: the transition speeds were not the same. *Reversing the direction of the transition changed something about the physics indicating a T-symmetry violation.*

Particles in a Rewinding Universe

Mathematically, particles in a T-transformation with a rewinding universe look like they went under a charge-parity inversion (left side of Figure 2-15).

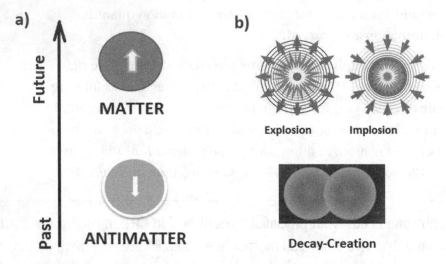

Figure 2-15. *T-transformations in CPT: (a) time reversal of antimatter as time reversed matter. (b) T-transformation by reversing all momentum and spin*

Matter that goes forward in time looks like antimatter going backward in time. This interpretation of antimatter as time reversed matter was first proposed by Swiss mathematician Ernst Stueckelberg in 1945; however, nowadays it is largely associated with Richard Feynman.

Tip   This interpretation is essential for Feynman's path integral approach to quantum mechanics and his ubiquitous Feynman diagrams. Furthermore, it is the inverse of a CP transformation, that is, the two undo each other. *This means that if CP is violated, then the simple time reversal is also violated.* This T-violation has been observed experimentally in the asymmetry between matter and antimatter.

CPT Is Safe: The 70-Year Rollercoaster Ride for the Symmetries of Nature

All physicists nowadays agree that quantum field theory is right as far as it goes. So much so that it is the most successful theory science has come up with. It is backed by an arsenal of experimental results that match the theory as closely as it can be expected.

All in all, we found out that the Stanford experiment pointed to a T-symmetry violation in this universe, but that is a good thing; remember that with CP symmetry also violated then T-symmetry violation saves CPT at the end. Thus far, in this decades-long rollercoaster, we have seen how symmetries have been broken one by one:

- First, parity was broken by the Cobalt-60 radioactive isotope experiment by Chien-Shiung Wu.

- Next, charge-parity was broken by the Neutral Kaon experiment by Cronin and Fitch.

- Finally, we were left with the choice of giving up on the symmetry of time itself in order to save CPT, but now, with T-symmetry looking like it is broken too, the CPT theorem is safe.

In the end, Feynman's antimatter time reversed clock works just fine, and our perfect mirrored universe works under a reversal of all three: charge, parity, and time.

Richard Feynman, the great pioneer of quantum field theory, passed away in 1988, at age 69 from liposarcoma, a rare form of cancer in soft tissue. His death was and always will be a terrible loss for the world of physics. I compare his death to the loss of Chopin or Mozart, like the master composer; he dedicated his life to creating masterful works, and in that journey he became one of the most original minds of his generation. Feynman became a legend in the scientific community not only for his groundbreaking work on the path integral approach to quantum mechanics and quantum field theory, but for his outrageous and dazzling method of teaching. His lectures in physics and his book *Six Easy Pieces* are mythical and earned him monikers such as *The Great Explainer* and *Physics Most Brilliant Teacher*. Near his death, he told a colleague that after teaching so much and meeting so many people, he may not be really gone. If his theories about quantum fields prove to be ultimately correct, then his genius will not be gone, just recycled, perhaps into the mind of a young child out there riding his first bike, looking at the stars, and wondering about the sea of truth out there waiting to be discovered. Long live Richard Feynman: The Great Explainer.

Exercises

Put your knowledge of QFT and QED to the test with this exercise set.

1. What do QFT and QED stand for? How does QED relate to QFT?

2. In one sentence, what does QED study?

3. What three ways of managing the complexity in QFT such as in the Dirac equation or Feynman path integral?

4. In one sentence define electron scattering. What law or equation is used to describe it in classical electrodynamics?

5. In classical electrodynamics, what formula is used to calculate the force between two stationary charged electrons?

6. Which of the following is true:

   a. In QFT electrons are vibrations in the electron field.

   b. Quantum fields permeate the entire universe.

   c. The electron field does not interact with the electromagnetic field.

   d. Vibrations in the electromagnetic field generate protons and quarks.

7. Fill in the blanks. Feynman diagrams are also _____ in disguise with each part of the diagram representing a chunk of the maths. In a Feynman diagram when two electrons travel toward each other, they may exchange a _____ _____.

8. Which of the following is true about perturbation theory:

   a. It is the art of drawing equations.

   b. If you have an impossible equation, find a similar equation that is solvable and make small changes to it so the result is similar.

   c. The more complicated the interaction, the less it contributes to the final probability amplitude.

   d. The probability of a particular interaction depends on the number of vertices of a Feynman diagram.

9. What is a loop interaction in a Feynman diagram? Give two examples.

10. Why may a loop interaction become infinite?

11. How does renormalization eliminate infinite loops in QFT?

12. Describe a pitfall of renormalization.

13. Which of the following is true about the Feynman path integral:

    a. There are infinite possible intermediate states in a quantum system evolving between two states.

    b. Impossible paths must not be considered.

    c. The probability of a path is impossible to calculate even with Feynman diagrams.

14. Draw the basic elements of a Feynman diagram as well as coordinates.

15. Draw the most probable interaction for electron scattering using a Feynman diagram. Explain what is going on. Don't use any formulas.

16. What is a virtual particle?

17. Give two weird characteristics of virtual particles.

18. How do Feynman diagrams simplify QFT calculations?

19. There is an interesting case in QED called *Bhabha scattering*, named after the father of the Indian nuclear program, physicist Homi J. Bhabha (1909–1966). Bhabha scattering is the electron-positron scattering process. In this exercise, you must draw the two most important Feynman diagrams for Bhabha scattering using a single virtual photon and two vertices each. Also, describe in a few words what goes on in each interaction.

    • Hint 1: Use two vertices with a single virtual photon for both cases. Note that both diagrams may seem to describe two very different events but must give the same result.

- Hint 2: Use the simple rules described in this chapter and remember the power of diagram rotation.

20. Draw all the possible four vertex diagrams for Bhabha scattering from the previous exercise. Ignore all self-energy diagrams (when an electron or positron emits and reabsorbs a photon). Explain what transpires on each diagram.

    Hint: Ignore all diagrams showing self-energy loops as shown in the following diagram.

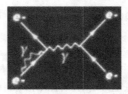

21. Fill in the blanks. A line dividing a square, triangle, or circle in the middle creates halves that are examples of _____ symmetry.

22. Briefly describe two types of symmetries in nature. Give an example of each.

23. Fill in the blanks. A parity transformation involves the flipping of___ ____. A true x-axis transformation equals a z-axis reflection plus a ____.

24. What does it mean to be P-symmetric?

25. In a mirror universe what properties get flipped and which ones do not?

26. True or false: The laws of physics should work the same in the mirror world even if the spatial axes get flipped.

27. Is parity symmetry (P-symmetry) violated in our universe? Why or why not?

28. In a few sentences describe CP symmetry.

29. Briefly describe the experiment that showed CP symmetry violation.

30. What is CPT symmetry? Why is it so important?

31. Give two examples of T-symmetry transformations.

## CHAPTER 3

Behold, the Qubit Revolution

At the heart of a quantum computer is the qubit, designed as the analog of the classical bit, the deterministic component at the heart of all electronics out there. Bits are physically constructed using a transistor. Transistors are tiny, at around 15 nanometers (nm) where 1 nm = 10^{-12}m; qubits, on the other hand, are big – a few meters tall. In this chapter, we look at the basic architecture of the qubit as designed by the pioneering IT companies in the field.

You will also learn that although qubits are mostly experimental and difficult to build, it doesn't mean that one can't be constructed with some optical tools and some ingenuity. Even if crude and primitive, a quantum gate can be built using refraction crystals, photon emitters, and a low budget. This chapter also explores superconducting loops as the main method for building qubits along with other popular designs and their relationship to each other. Each particular design has its own benefits and caveats when it comes to size, cost, and performance. So let's get started.

Your Friendly Neighborhood Quantum Computer

In the last few decades, quantum computers took the leap from the theoretical realm into the experimental one. And, although regular people cannot afford the 15 million USD for a shiny 2000 qubit D-Wave system, believe or not a very primitive quantum gate can be constructed using some understanding of optics and a sanguine knowledge of physics. In this section, we take the experimentalist plunge and build a crude yet effective quantum gate capable of entangling photons to produce a controlled-Z (CZ) gate. This gate is the hardest to build and demonstrates the quantum property of superposition. But before we look at the hardware required to build this gate, let's look at the theory behind it.

© Vladimir Silva 2024
V. Silva, *Quantum Computing by Practice*, https://doi.org/10.1007/978-1-4842-9991-3_3

In the golden age of quantum mechanics in the early 20th century, the great Austrian physicist Wolfgang Pauli (1900–1958) won the Nobel Prize for his remarkable research on the *Exclusion Principle*. In simple terms, this principle states that: *no two particles can occupy the same quantum state at the same time*. This may not sound like much to the un-initiated in atomic physics, but it was a stroke of genius by Pauli. As a matter of fact, Einstein in his letter of recommendation on behalf of Pauli to the Novel committee stated: *For his scientific research on discovering a new law of nature*.

Pauli's exclusion principle explains the stability of electron states in heavy elements with some physicists suggesting that Pauli's principle is responsible for the fact that ordinary bulk matter is stable and occupies volume. Results of this remarkable research are the Pauli spin matrices (X, Y, Z) commonly denoted by the Greek letter sigma (σ) thus:

$$\sigma_x = \begin{bmatrix} 0 & 1 \\ 1 & 0 \end{bmatrix} \sigma_y = \begin{bmatrix} 0 & -i \\ i & 0 \end{bmatrix} \sigma_z = \begin{bmatrix} 1 & 0 \\ 0 & -1 \end{bmatrix}$$

Pauli's matrices are 2x2 complex matrices where (i) denotes the imaginary number $\sqrt{-1}$.

Tip   In quantum mechanics the Pauli matrices are used to describe the spin of a particle when it interacts with an external electromagnetic field.

Pauli gates, in turn, can be used to construct more complex gates. For example, here is a 2 qubit controlled gate using the Z-gate. This one is more powerful, and capable of superposition and entanglement (see Figure 3-1).

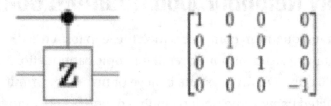

Figure 3-1. *Controlled Z-gate (CZ) shown in diagram and matrix modes. Note that all quantum gates have a corresponding matrix representation*

In Figure 3-1, the horizontal lines represent the qubits. The black dot is called the control, and it operates in the second qubit by performing the Z operation only when the first qubit is |1>, and otherwise leaves it unchanged.

Tip   In quantum mechanics, all gates are described using matrices and column vectors or kets |0>, |1> (more about this in the next chapter). It is the rich representation of matrices that gives the performance boost over classical computing.

Table 3-1 details the hardware components that can be used to build a rudimentary (CZ) quantum gate as well as their estimated cost. These components cannot be found at the local radio shack but can be purchased from eBay for cheap. They can also be purchased from any optics manufacturer.

Table 3-1. *Required hardware to build a photonic CZ quantum gate*

Item	Quantity	Description	Estimated cost (USD)
USHIO EXCIMER PHOTON SOURCE UER20H, UER 20 H 126 VA, USHIO EH0019	2	Single photon source	800
USHIO EXCIMER Photon Source Power Supply B0083	1	Photon source power supply	1350
Partially Polarizing Beam Splitters (PPBS)	3	An optical device that splits a beam of light in two	20

Let's look at these items in more detail:

- Single photon source: This is a light source that emits single particles of photons. Each source represents an input qubit to the CZ gate.

- Power supply: It would be great if we could plug in our photon source to the AC power outlet; however, this is not possible; we need a pricey custom power supply. In this case, I have used a device from Japanese optics manufacturer USHIO from eBay.

- Partially Polarizing Beam Splitter (PPBS): This is essentially a crystal with two specific properties perfect to study quantum mechanical effects:

  - It splits a beam of light in two.

  - It acts as an optical filter that lets light waves of a specific polarization pass through while blocking light waves of other polarizations.

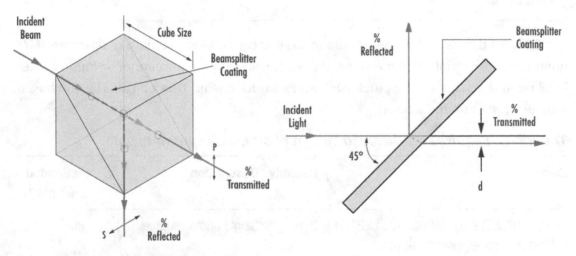

Figure 3-2. *Polarizing beam splitter types*

A beam splitter crystal comes in two flavors:

1. A cube constructed using two right triangle prisms: The hypotenuse of one prism is coated, and the two prisms are cemented together to form a cube (see left side of Figure 3-2). This model has the advantage of eliminating beam shift with equally reflected and transmitted optical paths. However, its heavy solid glass construction is more difficult and expensive to build in large sizes.

2. Linear plates: Made of a thin, flat glass plate coated on the first surface of the substrate. They are lightweight, relatively inexpensive, and easy to manufacture in larger sizes. However, their reflected and transmitted optical paths are of different lengths, and they produce a beam shift of transmitted light (right side of Figure 3-2).

Tip   Using optics to build a quantum gate is described in the *New Journal of Physics* by T Nagata et al.[1] As a matter of fact, our setup seeks to replicate Nagata's experiment more cheaply.

With the hardware from Table 3-1, we can replicate Nagata's pioneering work of building a quantum CZ gate using optics. His setup is described in Figure 3-3.

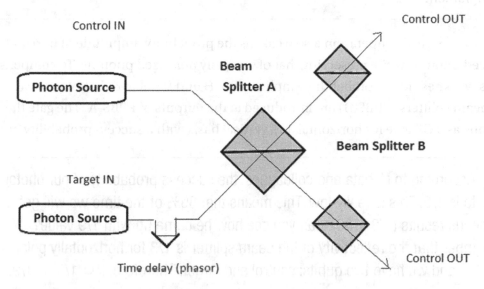

Figure 3-3.   CZ gate constructed using optics as described by Nagata and colleagues

The experimental setup in Figure 3-3 uses

- Two photon sources representing the control and target qubits of the gate.

- Both photons must enter the Partially Polarizing Beam Splitter (PPBS-A) simultaneously. The reflectivity of splitter-A is 1/3 for horizontally polarized light (H) and 1 for vertically polarized light (V). When the photons hit the splitter, a quantum mechanical effect (*two-photon quantum interference*) occurs **only for the horizontally polarized** photons (H).

- At the gate output, we count the cases in which single photons are emitted to both output modes: control and target.

---

[1] T Nagata et al. 2010 New J. Phys. 12 043053

The combination of two-photon interference and post-selection flips the phase or sign, that is, <H,H> becomes <-H, -H> of the input state only when both of the input photons are horizontally polarized.

Tip   The control-Z has the property that it does nothing to all input pairs except <H,H> which get mapped to <-H,-H>. This requires superposition of states and entanglement.

Furthermore, this operation also weakens the probability amplitude of horizontally polarized photons while preserving that of vertically polarized photons. To compensate for this weakness in the probability amplitudes of H and V polarized photons, two extra beam splitters (PPBS-B) are introduced at the outputs of PPBS-A. The gate then functions as a CZ gate for horizontal and vertical basis with a success probability of 1/9.

Tip   According to Nagata and colleagues, the success probability of our photonic CZ gate is 1/9. This is really low. This means that 89% of the time we will get erroneous results (8/9 *100). Can you see how he came up with 1/9 value? Remember that the reflectivity of the beam splitter is 1/3 for horizontally polarized light (H), and we have two qubits: control and target. Therefore, $P = 1/3^2 = 1/9$.

The high error probability of the photonic CZ gate is probably one of the reasons optical quantum gates have not taken off, with most of the big players in quantum choosing superconducting loops instead. As we'll see later on in this chapter, errors introduced by particle interactions with the external environment are a major source of headaches for quantum computation, and where most of the current research is geared toward. This is the age of the Noisy Intermediate Scale Quantum (NISQ) computer. As a matter of fact, most of Nagata's work on this experiment is concentrated on reducing the error rates produced by the beam splitters, but now let's take a closer look at the quantum effect at the heart of the gate: two-photon quantum interference.

Two-Photon Quantum Interference

Also known as the Hong-Ou-Mandel effect, it was demonstrated in 1987 by physicists Chung Ki Hong, Zhe Yu Ou, and Leonard Mandel at Rochester University. This effect occurs when two identical single photons enter a 1:1 beam splitter. The 1:1 means that the photon has a 50:50 chance of being reflected or transmitted. Under perfect conditions, the following applies:

- When the temporal overlap of the photons is perfect (they hit the splitter at the same time), the two photons will always exit the beam splitter together in the same output mode.

- If the photons become more distinguishable, the probability of them going to different detectors will increase.

- An interferometer, an instrument that merges two or more sources of light to create an interference pattern, which can be measured and analyzed, can be used to accurately measure bandwidth, path lengths, and timing.

Tip   Two-photon interference relies on the existence of photons and cannot be fully explained by classical optics. This effect provides the underlying physical mechanism for logic gates in linear optical quantum computation.

Thus, when the photons hit the splitter at the same instant, they can either be reflected or transmitted with four possibilities shown in Figure 3-4 and Table 3-2.

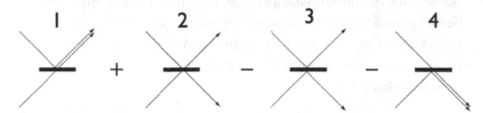

Figure 3-4. *Reflection/transmission possibilities in two-photon interference*

Table 3-2. *Possible outcomes of two-photon interference*

	Top	Bottom
1	Reflected	Transmitted
2	Transmitted	Transmitted
3	Reflected	Reflected
4	Transmitted	Reflected

Note   In physics, the so-called Feynman rule states that since the splitter does not record the state of the photons (is not a measuring device), then we must add all the possible states (paths) to calculate its probability amplitude. In quantum mechanics, the modulus squared of this value represents the probability of a specific outcome.

Furthermore,

- The reflection from the bottom side of the beam splitter introduces a relative phase shift of π, corresponding to a factor of -1 in the associated term in the superposition (this means options 3 and 4 must be subtracted from the probability amplitude). Note that this is required by the *reversibility* of the quantum states of the beam splitter.

- Because the two photons are identical, we cannot distinguish between the output states of possibilities 2 and 3; therefore, the minus sign of option 3 ensures that these two terms cancel each other. In wave mechanics this is called *destructive interference* (see right side Figure 3-5).

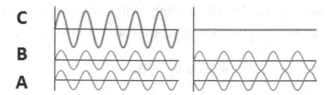

Figure 3-5. *Interference in wave mechanics. Add waves A and B to get constructive interference (left) or cancel each other (destructive) on the right*

Tip   Quantum mechanics asserts that the state of a quantum system is described by a *probability amplitude vector where all possibilities must sum to unity.* This idea was proposed by German physicist Max Born (1882–1970) who was instrumental in the development of quantum mechanics in its early days. In Born's interpretation, any transformation of a quantum state must preserve the length of the state vector. This implies that linear transformations between quantum states must be unitary and reversible (all states are described by probabilities with complex coefficients and must add to 1).

Mathematics Behind Photonic Interference

To understand what goes on behind two-photon interference, quantum mechanics relies on annihilation and creation operators. These operators provide a solution to the quantum harmonic oscillator problem using the algebra, and are widely used in many-particle systems and the study of photons:

- The *creation* operator $\hat{a}^\dagger$ (a-hat-dagger): *Increases the number of particles in a given state by one*, and it is the inverse of the *annihilation* operator $\hat{a}$ which decreases the number of particles by one.

- Consider two optical modes a and b (one for each photon) with creation/annihilation operators: $\hat{a}^\dagger$, $\hat{a}$ *and* $\hat{b}^\dagger$, $\hat{b}$.

Given a single photon state described using Dirac's ket-notation as $|1\rangle$, then the state of the two photons can be described using the operators above as

$$\hat{a}^\dagger \, \hat{b}^\dagger \, |0,0\rangle_{ab} = |1,1\rangle_{ab}$$

That is, apply the creation operator to the basis photon state for optical modes (a, b) to obtain its current state. Now, when the two photons hit the beam splitter, one on each side, the optical modes (a, b) undergo a unitary transformation into (c, d) with the creation and annihilation operators transforming accordingly:

$$\hat{a}^\dagger \rightarrow \frac{\hat{c}^\dagger + \hat{d}^\dagger}{\sqrt{2}} \; and \; \hat{b}^\dagger \rightarrow \frac{\hat{c}^\dagger - \hat{d}^\dagger}{\sqrt{2}}$$

The minus sign in the second operator comes from the unitary transformation performed in matrix form: $\begin{pmatrix} \hat{a} \\ \hat{b} \end{pmatrix} = \frac{1}{\sqrt{2}} \begin{pmatrix} 1 & 1 \\ 1 & -1 \end{pmatrix} \begin{pmatrix} \hat{c} \\ \hat{d} \end{pmatrix}$

Finally, the state of the two photons is shown in Figure 3-6.

$$|1,1\rangle_{ab} = \hat{a}^\dagger \, \hat{b}^\dagger \, |0,0\rangle_{ab} \rightarrow \frac{1}{2}(\hat{c}^\dagger + \hat{d}^\dagger)(\hat{c}^\dagger - \hat{d}^\dagger)|0,0\rangle_{cd} = \frac{|2,0\rangle_{cd} - |2,0\rangle_{cd}}{\sqrt{2}}$$

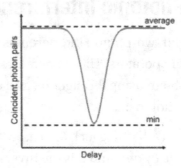

Figure 3-6. *State of the two photons when they enter the beam splitter along with their experimental signature*

Experimentally, two-photon interference can be observed using two photodetectors by monitoring the output modes (a, b) of the beam splitter: The coincidence rate of the detectors will drop to zero when the photons hit the splitter at the same time. This produces the supposed Hong-Ou-Mandel dip. Note that the coincidence count reaches zero when the two photons are perfectly identical in all properties (bottom of Figure 3-6).

Tip   *Ket-notation* $|1,1\rangle$ is fundamental in the mathematical description of quantum mechanics and was developed by the great English physicist Paul Dirac. A *ket* is essentially a *vector* of the states (s1, s2, …, sn) of a system or $|s1, s1, .., sn\rangle$. Note that, what goes inside the ket are labels: symbols, letters, numbers, or even words. For example, the basis states of a qubit are denoted by the kets $|0\rangle$, $|1\rangle$ where 0, 1 are labels (not the numbers 0, 1) – an unfortunate yet confusing accident. Additionally, since kets are vectors, they obey the usual rules of linear algebra. For example, $|A\rangle = |B\rangle + |C\rangle$.

Output States of the Control-Z Gate

In Nagata's gate when two photons enter beam splitter-A, quantum interference kicks in, then by using the creation/annihilation operators for the input modes (a,b), the unitary transformation U_A becomes

$$\begin{pmatrix} \hat{a}^\dagger \\ \hat{b}^\dagger \end{pmatrix} = U_A \begin{pmatrix} \hat{a} \\ \hat{b} \end{pmatrix} = \begin{pmatrix} \sqrt{R} & i\sqrt{1-R} \\ i\sqrt{1-R} & \sqrt{R} \end{pmatrix} \begin{pmatrix} \hat{a} \\ \hat{b} \end{pmatrix}$$

where (R) is the reflectivity of the splitter and (I) is the imaginary number. In contrast, beam splitter-B has four input modes when accounting for the vertical (V) and horizontal (H) polarizations (right side of Figure 3-7).

$$\begin{pmatrix} \hat{a}_H^\dagger \\ \hat{b}_H^\dagger \\ \hat{a}_V^\dagger \\ \hat{b}_V^\dagger \end{pmatrix} = U_B \begin{pmatrix} \hat{a}_H \\ \hat{b}_H \\ \hat{a}_V \\ \hat{b}_V \end{pmatrix}$$

$$U_B = \begin{pmatrix} U_A & 0 \\ 0 & e^{i\varphi} U_A \end{pmatrix} = \begin{pmatrix} \sqrt{RH} & i\sqrt{1-RH} & 0 & 0 \\ i\sqrt{1-R} & \sqrt{RH} & 0 & 0 \\ 0 & 0 & e^{i\varphi}\sqrt{Rv} & ie^{i\varphi}\sqrt{1-Rv} \\ 0 & 0 & ie^{i\varphi}\sqrt{1-Rv} & e^{i\varphi}\sqrt{Rv} \end{pmatrix}$$

where φ (Phi) is the phase difference between horizontally polarized light and vertically polarized light. In the ideal case, the parameters of PPBS-A are RH = 1/3, RV = 1, and φ = 0, so that the matrix U_B becomes

$$U_B = \begin{pmatrix} \sqrt{1/3} & i\sqrt{2/3} & 0 & 0 \\ i\sqrt{2/3} & \sqrt{1/3} & 0 & 0 \\ 0 & 0 & 1 & 0 \\ 0 & 0 & 0 & 1 \end{pmatrix}$$

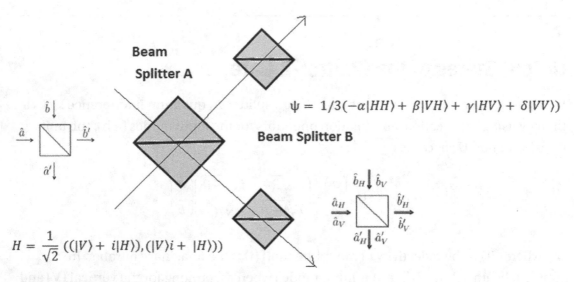

Figure 3-7. *Output states of the CZ gate*

- Beam splitter-A puts both photons in superposition. This is also known as a **Hadamard** transformation (a very useful transform/gate in quantum computation). The state of the system at this point for photon polarizations (H,V) is $H = \frac{1}{\sqrt{2}}\left(\left(|V\rangle + i|H\rangle\right), \left(|V\rangle i + |H\rangle\right)\right)$.

- When the photon cross beam splitter-B, four input modes are introduced for the vertical (V) and horizontal (H) polarizations with the final state of the system (Psi) $\psi = 1/3(-\alpha|HH\rangle + \beta|VH\rangle + \gamma|HV\rangle + \delta|VV\rangle)$.

The final state ψ is obtained by applying the unitary transformation U_B the four basis states for the photons with polarizations (H,V) using the creation operators with the ideal values for reflectivity ($R_H = 0$, $R_V = 2/3$). See Table 3-3.

Table 3-3. *Final states of the CZ gate*

Basis state	Resulting state
$U_B \mid H, H \rangle a, b$	$-\frac{1}{3} \mid H, H \rangle \hat{a}^\dagger \hat{b}^\dagger$
$U_B \mid H, V \rangle a, b$	$\frac{1}{3} \mid H, V \rangle \hat{a}^\dagger \hat{b}^\dagger$
$U_B \mid V, H \rangle a, b$	$\frac{1}{3} \mid V, H \rangle \hat{a}^\dagger \hat{b}^\dagger$
$U_B \mid V, V \rangle a, b$	$\frac{1}{3} \mid V, V \rangle \hat{a}^\dagger \hat{b}^\dagger$

Building quantum gates using linear optics seems easy when compared with other qubit designs such as superconducting loops which are complex and dangerous to build requiring cooling to almost absolute zero; the trade-off being significant reduction in error rates. Nevertheless, the race is on to build a high fidelity quantum gate. Who will win, nothing is set in stone when it comes to gate design.

Lowering Error Rates

Nagata and colleagues conclude their research with a deep analysis of the error sources for the photonic CZ gate, looking for clues on how to lower the error rate to achieve high fidelity. They identified the following sources of error:

- Optical birefringence error $\delta\varphi$: Birefringence is formally defined as the double refraction of light in a transparent, molecularly ordered material, which is manifested by the existence of orientation-dependent differences in the refractive index. All transparent solids such as glass, crystal polymers, even table salt produce optical birefringence.

- Mode mismatch error $\delta\varepsilon$: This error is introduced by the two optical modes (a, b) of each photon and their creation/annihilation operators.

- Reflectivity error for the horizontally polarized photons δR_H: This value has an effect in the accuracy of measurements.

- Reflectivity error for the vertically polarized photons δR_V.

The requirements for achieving a high-fidelity gate have been investigated by manipulating these four error sources with some potential innovations:

- High-reflectivity mirrors: These are ultra-high-tech crystals where the sum of the transmittance and the loss $(\sim 1 - R) < 10^{-5}$. They can be effective in reducing reflectivity errors as well as optical birefringence.

- Gates using fibers or waveguides to minimize the spatial mode mismatch $\delta \varepsilon$.

- Single-photon sources with small temporal jitter can also minimize temporal mode mismatch.

Superconducting Loops vs. Linear Optics

A gate made using linear optics is relatively simple, and inexpensive to build, but has high error rates. On the other hand, a superconducting loop gate is more complex and requires near absolute zero temperature. Yet they have a much lower error rate than photonics.

In this section, we look at the inner workings of these superconductor devices which a few years ago only existed in theoretical physics but now, slowly yet steadily, are becoming a reality, and soon will be all over the data center.

Superconducting Loops

When an electric current passes through a conductor, some of the energy is lost in the form of heat and light. This is called resistance, and it depends on the type of material; some metals like copper or gold are great conductors of electricity and have low resistance. Scientists discovered that the colder the material is the better conductor of electricity it becomes (with lower temperature comes less resistance). The problem was, no matter how cold the material gets, it will always show a level of resistance.

Note   In 1911, scientists discovered that when cooling down mercury to 4.2 degrees Kelvin (above absolute zero), its resistance becomes zero. This Eureka moment leads to the discovery of the superconductor, a material that has zero electrical resistance at very low temperatures.

Since then, many other superconducting materials have been found: aluminum, gallium, niobium, and others which show zero resistance at a critical temperature (see Table 3-4). The great thing about superconductors is that electricity flows without any loss, so a current in a closed loop can theoretically flow forever. As a matter of fact, this principle has been proved experimentally when scientists were able to maintain electricity flowing over superconducting rings for years.

Tip   In a qubit made of a superconductor loop, a current oscillates back and forth around a loop. A microwave is injected which excites the current into a superposition of states.

Table 3-4. Common superconductors and their critical temperatures

Material	Temperature (K)	Details
Mercury	4.2	First superconductor discovered in 1911 by Heike Kamerlingh Onnes
Lead	7	Found to superconduct in 1941
Niobium-nitride	16	A compound (alloy) of niobium and nitrogen commonly used in infrared light detectors
Niobium-titanium	16	Also known as a type II superconductor commonly used in industrial wires for superconducting magnets
Ceramics consisting of mercury, barium, calcium, copper, and oxygen	133	These are called high-temperature superconductors and are very desirable because they can be cooled with cheaper materials such as liquid nitrogen instead of the more expensive liquid helium

Breaking Out of the Lab: IBM-Q Qubit Design

Figure 3-8 shows a scaled down diagram of the basic components of a qubit using superconducting loops by the IBM-Q platform[2], as well as the temperature in Kelvin degrees. An important component of this design is the Qubit Signal Amplifier.

Qubit Signal Amplifier

This is a quantum mechanical signal booster used to read the qubit signal. Optical amplifiers are important in communications and laser physics. For example, they are widely used as repeaters in long distance fiber-optic cables to carry digital telecommunications.

Tip   There are two main properties of the quantum amplifier: its amplification coefficient and its uncertainty (noise). These are closely dependent: the higher the amplification coefficient, the higher the noise. Quantum systems are extremely sensitive to environmental noise.

For a qubit to do its job, we need the ability to read its quantum state with high fidelity in real time. To this purpose, a signal amplifier of a special kind can be used: quantum-limited amplifier (QLA). QLA devices use a low-noise microwave to probe the quantum state of the qubit while minimizing environmental noise. Nevertheless, this low noise comes at a price: The temperatures must be low, really low, colder than any place on Earth or even outer space (just a few notches above absolute zero). In Figure 3-8, QLAs are used in two places: on the top chamber of the qubit block at around 4K and in the inner chamber (where the superconducting loop lives) at a super frosty 10 millikelvin (10mK).

---

[2] IBM Quantum systems overview. Available online at https://quantum-computing.ibm.com/docs/manage/backends/

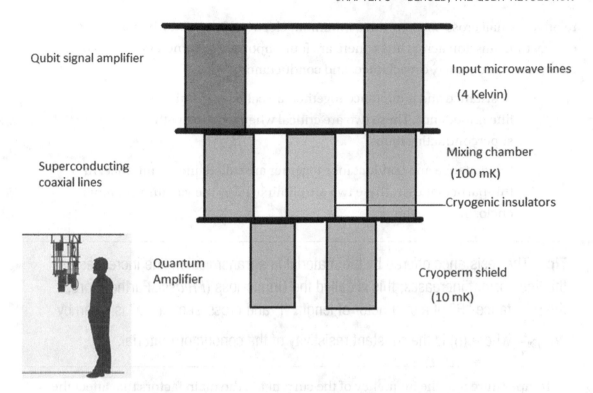

Figure 3-8. *Rough draft of the basic qubit design used by IBM-Q*

Note   To achieve these ultra-low temperatures, the qubit blocks are encased in a special cylindrical-shaped cryogenic chamber called a dilution fridge. The chamber has three temperature stages at 4K, 100mK, and 10mK with microwave signals carried by input and output chains designed to minimize the qubit exposure to noise. These include some microwave components, such as microwave and coaxial lines, cryogenic insulators, and shields.

Microwave and Superconducting Coaxial Lines

These are superconducting cables used to raise the overall current carrying capability and are widely used in cryogenic applications for the medical, scientific, and aerospace industries. They are useful because of their ability to carry very large currents in a relatively small cross-section with a minimum electrical loss, and when it comes to energy transmission across lines, there are four important parameters to consider: resistance, inductance, capacitance, and conductance.

- Resistance and inductance together are called transmission line impedance. These two are critical when working with superconducting loops.

- Capacitance and conductance together are called line admittance. In this particular case, these two are eliminated by the vacuum chamber enclosing the qubit.

Tip   The resistance offered by the material in a transmission line increases as the line current increases; this is called the Ohmic-loss (I^2Rloss). Furthermore, the resistance "R" of a conductor of length "l" and cross-section "a" is given by

$R = \rho\dfrac{l}{a}$ where (ρ) is the constant resistivity of the conductor material.

Temperature and the frequency of the current are the main factors that affect the resistance of a line: It is important to remember that the resistance of a conductor varies linearly with the change in temperature, whereas if the frequency of the current increases, the current density toward the surface of the conductor also increases. On the other hand, if the frequency decreases, the current tends to flow toward the center of the conductor (the more the current flows toward the surface, the less it flows toward the center, a property that is known as the skin effect).

This change in frequency gives rise to the concept of inductance (see Figure 3-9). In an AC transmission line, the current flows sinusoidally. This current induces a magnetic field perpendicular to the electric field, which also varies sinusoidally (an effect known as Faraday's law).

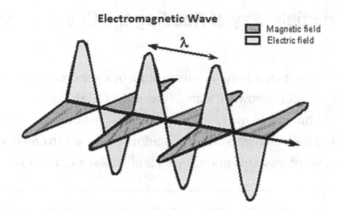

Figure 3-9. *Inductance or current flow in a transmission line*

This varying magnetic field induces an electro-magnetic-field (EMF) into the conductor which flows in the opposite direction of the initial current. This EMF flow in the opposite direction is described by the parameter known as inductance, which is the property to oppose the shift in the current.

Note   Superconductors reduce resistance and inductance to almost zero by lowering the temperature to a critical value and using special metals (mercury and lead) or ceramic alloys (niobium and titanium). This magic cocktail allows the transfer of high voltages with minimal loss/heat over tiny wires

---

Cryogenic Insulators and Shields

Their job is to provide thermal insulation while maintaining the best radiant energy barrier available for cryogenic containers and vacuum pipes. In other words, they prevent the toxic liquid nitrogen from escaping. All these components come together to tackle the big picture challenge: reading the qubit state with minimal noise, a difficult task commonly known as the *readout problem of superconducting qubits*.

The Non-destructible Way of Reading the Quantum State of a Qubit

The current and best way to read the quantum state of a superconductor is to probe its readout pulse with a weak microwave signal (around 7 GHz), that is, probe the signal with few microwave photons. However, this is not feasible at room temperatures without significantly boosting their energy, since the readout pulse is extremely weak (in the order of 10^{-16} Watts (W) or less; this adds appreciable noise to the amplified output.

Tip   The device used to probe the readout pulse is called a *Low Noise Amplifier (LNA)*. This is a state-of-the-art, high-gain, and low-noise semiconductor-based amplifier to boost the signal. LNAs by themselves are not enough to read the qubit state as they add about 1000 times as much noise as the quantum signal itself (top of Figure 3-10).

Engineers at IBM-Q solved the readout problem by adding a device called *Quantum Limited Amplifier* (QLA) as the first-stage amplification in the chain (see bottom of Figure 3-10).

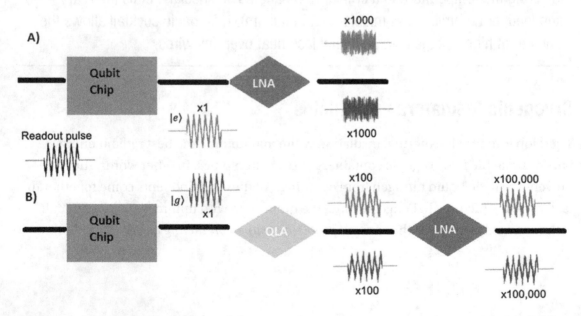

Figure 3-10. *Qubit signal amplification chain using LNAs and QLAs with a high signal-to-noise ratio (above) and low signal-to-noise (below)*

QLAs add only the minimum amount of noise required by quantum mechanics to the input signal (equal to the ambient quantum noise or half a photon at the signal frequency).

Tip   In the ideal case, the signal-to-noise ratio at the output of the QLA is only degraded by a factor of two (since the added noise and input noise are equal). QLAs work because the noise performance of a chain of microwave amplifiers is primarily set by the noise characteristics of the first amplification stage.

In Figure 3-10 a readout pulse enters the quantum chip (superconducting loop) and comes out as either ground $|g\rangle$ or excited $|e\rangle$; to probe the signal, an amplification chain is created using either LNAs or QLAs.

- By using only an LNA, the signal is blurred by a factor of 1000 and very difficult to resolve (due to the high signal-to-noise ratio).

- By using a QLA as the first amplification stage, the phase of the amplified readout signal at the output of the chain is easy to resolve as the ratio of the signal-to-noise is similar to that of the signal leaving the qubit chip.

To determine the final state of the qubit, researchers at IBM-Q use a metric called *readout fidelity*[3].

---

[3] Baleegh Abdo. Rising above the noise: quantum-limited amplifiers empower the readout of IBM Quantum systems. Available online at `www.ibm.com/blogs/research/2020/01/quantum-limited-amplifiers/`

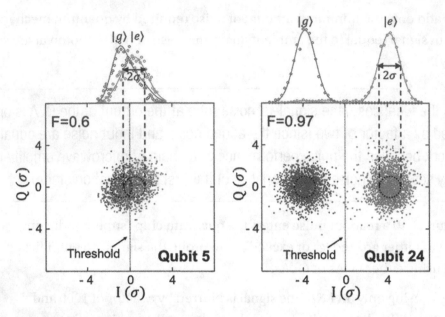

Figure 3-11. *Readout fidelity histogram of 2 qubits as reported by Abdo and colleagues from IBM research showing low (noisy) fidelity (left) and high (accurate) fidelity (right)*

Tip *Readout fidelity* represents the probability of a successful determination of the qubit state. This metric is encoded as a real number between 0.5 and 1, where 1 means we know the qubit state with certainty (no readout error) and 0.5 means failure to read the qubit state (readout error).

Figure 3-11 shows a histogram plot of the measurements of two qubit readout signals in the ground $|g\rangle$ (blue) and excited $|e\rangle$ (red) states. The diagram explains how readout fidelity is calculated: it is the overlap or separation of the ground or excited signals. The dots represent readout data points of I and Q, microwave signals; on top we see the Gaussian distributions of mean values (representing the signal strength) and the standard deviations (representing the noise level). On the left, both signals overlap with a fidelity of 0.6 (readout error); on the right, the signals are clearly distinguishable (fidelity = 0.97 – no error).

This readout fidelity measurement is the key to determine the qubit state in real time. In general, the separation between the signals depends on the strength of the probe signal (the number of photons used in the readout signal and the total gain of the output chain), while the standard deviation (measurement noise) depends on the noise in the output chain and the averaging time.

Note   Readout fidelity above 0.9 taken in less than 1 microsecond is highly desirable, and it is possible thanks to QLAs.

Thus, researchers at IBM-Q are hard at work enhancing the performance of the readout fidelity of their quantum processors with a twofold goal: tighter distributions in the readout fidelities and better control over time. This pioneering work has created quantum processors ranging from 5 to 53 qubits by the time of this writing. Furthermore, their plans for scaling are ambitious: A 1000 qubit processor by 2023[4] with an ultimate goal of millions of qubits.

Pros and Cons of Superconductor Loops

The race between optics and superconductors to build the ultimate quantum gate continues. Nevertheless, it is easy to see why big organizations have chosen superconducting loops. Here is a list of the most critical factors to be considered when building a high fidelity qubit:

- Error level: This is the probability of success or obtaining a valid result. For the photonic quantum gate described by Nagata, the success rate is 1/9 (11%). On the other hand, superconducting loops have success rates above 99%.

---

[4] Jay Gambetta. IBM's Roadmap for Scaling Quantum Technology. Available online at www.ibm.com/blogs/research/2020/09/ibm-quantum-roadmap/

- Longevity: This is the minimum amount of time a superposition of states can be kept. A very important metric, as this bizarre quantum mechanical effect is what gives quantum the performance edge over classical computing: Qubits exist in 0/1 superposition of states with a given probability, providing more paths to find a solution to the problem at hand.

- Cost: The average Joe may not be able to afford the cryogenic insulation required by superconductors, so physics labs at colleges may choose linear optics instead.

With this in mind, here are the pros vs. cons of superconductor qubits:

Advantages

- Low error levels (around 99.4% logic success rates).

- Fast, built on existing materials. Although very expensive and complex.

- Decent number of entangled qubits (around 9) capable of performing a 2-qubit operation.

Disadvantages

- Low longevity: 0.00005s. This is really low: superposition of states can be kept for at most 50 millionths of a second.

- Must be kept very cold, at a super frosty -271C. This is colder than space, just a few notches above absolute zero.

Lucky for us, optics or superconductors are not the only choices when building a high performance qubit. There are many options out there, with more popping out all the time.

The Many Flavors of the Qubit

The great physicist Richard Feynman once theorized that the bizarre properties of quantum mechanics could be used to construct an information processing device. As it turns out, any two-state (or two-level) quantum-mechanical system can be used to represent a qubit. The simplest examples of such states include

- The two level spin of the electron: Spin up and spin down (commonly used in superconducting loops)

- Single photon polarization: Vertical polarization and horizontal polarization (used in linear optics)

More esoteric two-state quantum systems include

- Electron charge: No electron, one electron.

- Nuclear spin via magnetic resonance: Where nuclei kept in a strong constant magnetic field produces an electromagnetic signal when perturbed by a weak oscillating magnetic field (up or down).

- Optical lattices: It uses laser beams to create periodic polarization patterns with states: up or down.

- Quantum dots: These are theoretical, tiny semiconductor particles nanometers in size that differ from larger particles due to quantum mechanics: when illuminated by UV light, an electron in the quantum dot can be excited to a state of higher energy (up). The excited electron can drop back into the basis state (down) releasing its energy by the emission of light.

All in all, there are plenty of two-state quantum mechanical systems that can be used to build a high fidelity qubit. Table 3-5 lists some of them, not counting the ones we have seen so far: superconducting loops and linear optics.

Table 3-5. *Some of the common qubit designs out there*

Ion Traps

Pros

- High longevity: Experts claim that trapped ions can hold entanglement for up to 1000s which is pretty good compared to superconductor loops (0.00005s)
- Better success rates (99.9%) than superconductors (99.4%). Not much but still
- Highest number so far (14) of entangled qubits capable of performing a 2-qubit operation

Cons

- Slow operation. Requires lots of lasers

An ion trap is a technique that uses a combination of electric or magnetic fields to capture charged particles (ions) in a system isolated from the external environment. Lasers are applied to couple qubit states for single operations or coupling between the internal states and the external motional states for entanglement

Scalability

Ion traps seek to realize the dream of large scale universal quantum computing by scaling with arrays of ion traps. This technique is also capable of building large entangled states via photon connected networks of remotely entangled ion chains, or combinations of these two

Silicon Quantum Dots

Pros

- Stable, built on existing semiconductor materials
- Better longevity than superconductor loops 0.03s

Cons

- Low number of entangled qubits (2) capable of performing a 2-qubit operation
- Lower success rate than superconductor loops or trapped ions, but still high at around 99%

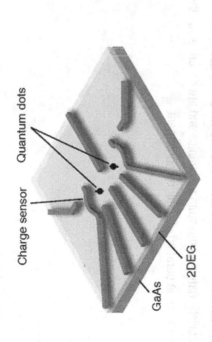

Charge sensor   Quantum dots

GaAs

2DEG

In a silicon quantum dot, electrons are confined vertically to the ground state of a quantum gallium arsenide (GaAs) forming a two-dimensional electron gas (2DEG). This two-dimensional electron gas is free to move in two dimensions, but tightly confined in the third. This tight confinement leads to quantized energy levels for motion in the third direction which are used to describe the state of a qubit

Note: 2DEG gases are commonly used in transistors made from semiconductors and can also exhibit quantum effects when conductance becomes quantized at low temperatures and strong magnetic fields (see the Hall effect)[5]

(continued)

[5] Hall, E. H. (1879). "On a New Action of the Magnet on Electric Currents." American Journal of Mathematics. JSTOR. 2 (3): 287. doi:10.2307/2369245

Table 3-5. *(continued)*

Topological Qubits

Pros

• Stable, error-free (longevity doesn't apply). This is incredible, if topological qubits could be realized, all the headaches of current quantum processors would be eliminated

Cons

• Purely theoretical at this point, although recent experiments indicate these elements may be created in the real world using semiconductors made of gallium arsenide at a temperature near absolute zero and exposed to strong magnetic fields

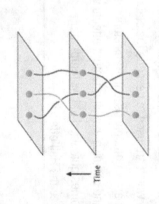

A topological qubit uses two-dimensional quasiparticles called *anyons* whose paths pass around one another to form braids in a three-dimensional space-time. These braids form the logic gates that make up the computer. Topological qubits seek to eliminate the error levels characteristic of quantum computers which are due to the probabilistic nature of quantum mechanics and are described by the longevity or the duration of qubit entanglement

Diamond Vacancies

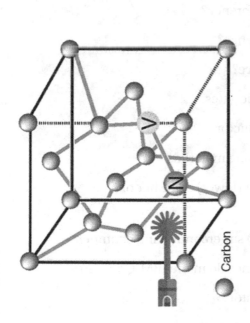

Carbon

Pros

- High longevity: 10s
- High success rate: 99.2%
- Decent number of entangled qubits (6) capable of performing a 2-qubit operation
- Qubits operate at room temperature

Cons

- Small number of vacancies in surface materials: about 2%
- Difficult to entangle

Diamond vacancies are meant to solve the perennial problem of reading information out of qubits *(measurement problem)* in a simple way. Remember that measuring the qubit state collapses its wave function and should be the very last thing in the execution chain. Now, because diamonds are natural light emitters, the light particles emitted preserve the superposition of states; thus, they could move information between quantum computing devices, a feat that would be considered the holy grail of quantum computation. Best of all, they work at room temperature; there is no need to cool things down to -272C degrees! One pitfall of diamond vacancies is that only about 2 percent of the surface of a natural diamond has them. Nevertheless, researchers are developing processes for blasting the diamond with beams of electrons to produce more vacancies

Quantum computers have come a long way since the days of Richard Feynman with some of the world's biggest companies looking to get in the game. Right now superconductor loops are leading the pack. However, there are amazing new designs, such as ion traps or diamond vacancies, seeking to realize the dream of large scale quantum computing.

Exercises

Here is a set of easy exercises to review the main concepts in this chapter. Answers are provided in the appendix.

1. Write down the matrix representation of the Pauli spin matrices X, Y, Z.

2. What is the matrix representation of a control-Z gate?

3. Briefly describe two types of bean splitter.

4. Give one reason why optical quantum gates have not taken off.

5. Briefly state the Feynman rule in physics.

6. What is two-photon quantum interference?

7. What are two types of wave interference?

8. What is destructive wave interference?

9. Give two sources of error in photonic gates.

10. In one sentence define a superconductor.

11. Give two examples of superconducting materials.

12. What is the temperature of the inner (lower) chamber of a superconducting qubit?

13. List three components of the IBM-Q superconducting chamber.

14. What do superconducting coaxial lines do in the IBM-Q design?

15. What is the job of a cryogenic insulator?

16. What is a non-destructible way of reading the quantum state of a qubit?

17. What are two devices used to probe the readout pulse in IBM-Q?

18. What is readout fidelity?

19. List three factors to be considered when building a high fidelity qubit.

20. What is the Born rule?

21. List three types (designs) of qubit systems.

## CHAPTER 4

Enter IBM Quantum: A One-of-a-Kind Platform for Quantum Computing in the Cloud

In this chapter we take a look at quantum computing in the cloud with IBM Q: the first platform of its kind. The chapter starts with an overview of the composer, the web console used to visually create circuits, submit experiments, explore hardware devices, and more. Next, you will learn how to create your first experiment and submit it to the simulator or real quantum device. IBM Quantum features a powerful REST API to control the life cycle of the experiment, and this chapter will show you how with detailed descriptions of the endpoints and request parameters. Finally, the chapter ends with a series of exercises to put your REST API skills to the test. Let's get started.

IBM has certainly taken an early lead in the race for quantum computing in the cloud. They came up with a powerful platform to run experiments remotely called IBM Quantum. Let's see how this platform has the power to transform the status quo.

© Vladimir Silva 2024
V. Silva, *Quantum Computing by Practice*, https://doi.org/10.1007/978-1-4842-9991-3_4

Getting Your Feet Wet with IBM Quantum

This is IBM's platform for quantum computing; let's take a look – follow the steps outlined here (All Reprints Courtesy of International Business Machines Corporation, © International Business Machines Corporation):

- Create an account in `https://quantum-computing.ibm.com/`. You will need an email; wait for the approval and confirm.

- Login to the web console and navigate to the composer tab by clicking the *Learning* menu from the application switcher in the top right of the main menu.

Quantum composer

The composer is the visual tool used to create your quantum circuits. At the top, it shows the experiment histogram with qubits available for use (see Figure 4-1).

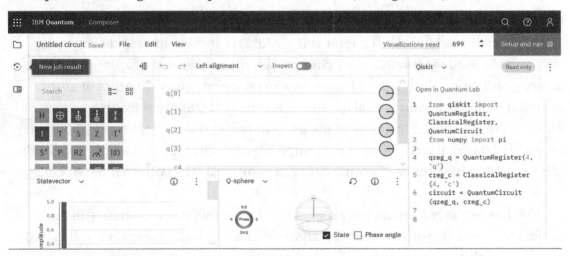

Figure 4-1. *Experiment histogram from the composer*

- On the right side, we see the histogram with 4 qubits available for use. They are all initialized to the ground state |0>. This is where you place your quantum gates.

- On the left side, we have the quantum gates. Drag gates into the histogram location of a specific qubit to start building a circuit.

Let's look at the gates and their meaning.

Quantum Gates

The quantum gates supported by IBM Quantum are described in Table 4-1.

Table 4-1. *Quantum gates for IBM Quantum*

Gate	Description
Pauli X ⊕	It rotates the qubit 180 degrees in the X-axis. Maps |0> to |1> and |1> to |0>. Also known as the bit flip or NOT gate. It is represented by the matrix: $$X = \begin{bmatrix} 0 & 1 \\ 1 & 0 \end{bmatrix}$$
Pauli Y Y	It rotates around the Y-axis of the Bloch sphere by π radians It is represented by the Pauli matrix: $$Y = \begin{bmatrix} 0 & -i \\ -i & 0 \end{bmatrix}$$ where $i = \sqrt{-1}$ is known as the imaginary unit
Pauli Z Z	It rotates around the Z-axis of the Bloch sphere by π radians It is represented by the Pauli matrix: $$Z = \begin{bmatrix} 1 & 0 \\ 0 & -1 \end{bmatrix}$$
Hadamard H	It represents a rotation of π on the axis $(X + Z)/\sqrt{2}$. In other words, it maps the states: • |0> to $(\|0> + \|1>)/\sqrt{2}$ • |1> to $(\|0> - \|1>)/\sqrt{2}$ This gate is required to make superpositions
Phase $\sqrt{Z}$ S	It has the property that it maps X→Y and Z→Z. This gate extends H to make complex superpositions
Transposed Conjugate of S S†	It maps X→-Y, and Z→Z

(continued)

Table 4-1. (*continued*)

Gate	Description
Controlled NOT (CNOT)	This is a two qubit gate that flips the target qubit (applies Pauli X) if the control is in state 1. This gate is required to generate entanglement
Phase $\sqrt{S}$	The $\sqrt{S}$ gate performs a halfway rotation of a two-qubit swap. It is universal such that any gate can be constructed from only sqrt(swap) and single qubit gates. It is represented by the matrix: $$\sqrt{S} = \begin{bmatrix} 1 & 0 & 0 & 0 \\ 0 & 1/2(1+i) & 1/2(1-i) & 0 \\ 0 & 1/2(1-i) & 1/2(1+i) & 0 \\ 0 & 0 & 0 & 1 \end{bmatrix}$$
Transposed Conjugate of T or Dagger-T	Represented by the matrix: $$\sqrt{S} = \begin{bmatrix} 1 & 0 & 0 & 0 \\ 0 & 1/2(1-i) & 1/2(1+i) & 0 \\ 0 & 1/2(1+i) & 1/2(1-i) & 0 \\ 0 & 0 & 0 & 1 \end{bmatrix}$$
Barrier	It prevents transformations across its source line
Measurement	The measurement gate takes a qubit in a superposition of states as input and spits either a 0 or 1. There is a probability of a 0 or 1 as output depending on the original state of the qubit. Always remember that measurement should be the last thing done in the circuit
Conditional	Conditionally apply a quantum operation
Identity	The identity gate performs an idle operation on the qubit for a time equal to one unit of time

You can drag gates from the left side of the composer to create a circuit, or if you prefer to write code, you can switch to the editor mode as shown in Figure 4-2.

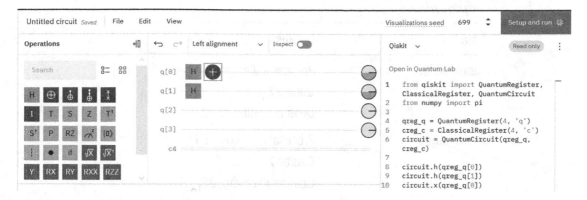

Figure 4-2. *Experiment editor in QASM editor mode*

Tip   The editor gives you two language choices: Qiskit or QASM (Quantum Assembly). At the end, your circuits or Python code will be translated into QASM for submission into a real device.

Now let's take a look at the various quantum processors available for use

Quantum Backends Available for Use

There are many quantum processors available for experimentation under the open (free) plan (from the main menu click Compute Resources). Table 4-2 shows a partial list ranked by the number of qubits according to the IBM Quantum backend information site.[1]

---

[1] IBM Quantum backend information available at `https://quantum-computing.ibm.com/lab/docs/iql/manage/account/ibmq`

123

Table 4-2. *Partial list of quantum backends available for IBM Quantum users under the open plan*

Name	Details
ibm_perth	Processor: Falcon r5.11H
	Qubits: 7
	Quantum Volume: 32
ibm_lagos	Processor: Falcon r5.11H
	Qubits: 7
	Quantum Volume: 32
ibm_nairobi	Processor: Falcon r5.11H
	Qubits: 7
	Quantum Volume: 32

Here is a little secret: there is a very interesting way to get an updated list of available machines in real time using the excellent REST API. This API is described in more detail in the "Remote Access" section in this chapter, but for now let's demonstrate how to obtain an always up-to-date list of backends using the *Available Backend List* REST endpoint: `https://api-qcon.quantum-computing.ibm.com/api/Backends?access_token=ACCESS-TOKEN`.

Tip   To obtain an access token, see the section "Authentication via API Token" under "Remote Access" of this chapter. Note that an API token is not the same as an access token. API tokens are used for authentication. Access tokens are used to invoke operations using the REST API.

The URL in the previous paragraph returns a list of quantum processors in JSON format. This is what it looks like by the time of this writing. Note that your results may be different:

Listing 4-1. HTTP response from the backend information REST API call

```
[{
  "name": " ibm_lagos",
  "version": "1",
  "status": "on",
  "serialNumber": "Real5Qv2",
  "description": "5 transmon bowtie",
  "basisGates": "u1,u2,u3,cx,id",
  "onlineDate": "2017-01-10T12:00:00.000Z",
  "chipName": "Sparrow",
  "id": "28147a578bdc88ec8087af46ede526e1",
  "topologyId": "250e969c6b9e68aa2a045ffbceb3ac33",
  "url": "https://ibm.biz/qiskit-ibmqx2",
  "simulator": false,
  "nQubits": 7,
  "couplingMap": [
    [0, 1],
    [0, 2],
    [1, 2],
    [3, 2],
    [3, 4],
    [4, 2]
  ]
}, {
  "name": "ibm_nairobi ",
  "version": "1",
  "status": "on",
  "serialNumber": "ibmqx5",
  "description": "16 transmon 2x8 ladder",
  "basisGates": "u1,u2,u3,cx,id",
  "onlineDate": "2017-09-21T11:00:00.000Z",
  "chipName": "Albatross",
  "id": "f451527ae7b9c9998e7addf1067c0df4",
  "topologyId": "ad8b182a0653f51dfbd5d66c33fd08c7",
  "url": "https://ibm.biz/qiskit-ibmqx5",
```

```
    "simulator": false,
    "nQubits": 7,
    "couplingMap": [
      [1, 0],
      ...
      [15, 14]
    ]
  },
  {
    "name": "ibmqx_hpc_qasm_simulator",
    "status": "on",
    "serialNumber": "hpc-simulator",
    "basisGates": "u1,u2,u3,cx,id",
    "onlineDate": "2017-12-09T12:00:00.000Z",
    "id": "084e8de73c4d16330550c34cf97de3f2",
    "topologyId": "7ca1eda6c4bff274c38d1fe66c449dff",
    "simulator": true,
    "nQubits": 32,
    "couplingMap": "all-to-all"
  },
  {
    "name": "ibmqx_qasm_simulator",
    "status": "on",
    "description": "online qasm simulator",
    "basisGates": "u1,u2,u3,cx,id",
    "id": "18da019106bf6b5a55e0ef932763a670",
    "topologyId": "250e969c6b9e68aa2a045ffbceb3ac33",
    "simulator": true,
    "nQubits": 24,
    "couplingMap": "all-to-all"
  }]
```

Listing 4-1 shows that there is a lot of extra interesting information about the structural layout of these machines:

- Extra processors and simulators:

  - There are a few remote simulators available for use (ibmq_qasm_simualtor, simulator_mps, ibmq_qasm_simualtor). This information can come in handy when testing complex circuits: more simulators are always a good thing.

  - Rumors of a 475 qubit processor have been swirling around for some time. There is even talk of an upcoming 1000 qubit processor by the end of 2023, but don't get excited just yet; this machine is only available for corporate customers. IBM has put an ambitious road map to reach a million qubits in the future. Check it out at www.ibm.com/quantum/roadmap.

- Besides the usual information such as machine name, version, status, number of qubits, and others. There are some terms we should be familiarized with:

  - basisGates: These are the physical qubit gates of the processor. They are the foundation under which more complex logical gates can be constructed. Most of the processors in the list use u1,u2,u3,cx,id.

    - Gates u1, u2, u3 are called *partial NOT gates* and perform rotations on axes X, Y, Z by theta, phi, or lambda radians of a qubit.

    - Cx is called the *controlled NOT* gate (CNOT or CX). It acts on 2 qubits and performs the NOT operation on the second qubit only when the first qubit is |1>, and otherwise leaves it unchanged.

    - Id is the identity gate that performs an idle operation on a qubit for one unit of time.

- couplingMap: The coupling map defines interactions between individual qubits while retaining quantum coherence. Qubit coupling is important, to simplify quantum circuitry and allow the system to be broken up into smaller units.

Now back to the composer for our first quantum circuit.

Entanglement: Bell and GHZ States

Entanglement experiments are used to demonstrate the weirdness of quantum mechanics and come in two flavors:

- Bell states: They demonstrate that physics is not described by local reality. This is what Einstein called *spooky action at a distance* (2-qubit entanglement).

- GHZ states: Even stranger than Bell states (named after their creators: Greenberger-Horne-Zeilinger), they describe 3 qubit entanglement.

Let's look at them in more detail.

Two Qubit Entanglement with Bell States

Bell states are the experimental test of the famous Bell inequalities. In 1964 Irish Physicist John Bell proposed a way to put quantum entanglement (spooky action at a distance) to the test. He came up with a set of inequalities that have become incredibly important in the physics community. This set of inequalities is known as Bell's theorem and it goes something like this.

Consider photon polarization (when light oscillates in a specific plane) at three different angles A = 0, B = 120, C = 240. Realism says that a photon has definite simultaneous values for these three polarization settings, and they must correspond to the eight cases shown in Table 4-3.

Table 4-3. *Permutations for photon polarizations at three angles*

Count	A(0)	B(120)	C(240)	[AB]	[BC]	[AC]	Sum	Average
1	A+	B+	C+	1(++)	1(++)	1(++)	3	1
2	A+	B+	C-	1(++)	0	0	1	1/3
3	A+	B-	C+	0	0	1(++)	1	1/3
4	A+	B-	C-	0	1(--)	0	1	1/3
5	A-	B+	C+	0	1(++)	0	1	1/3
6	A-	B+	C-	0	0	1(--)	1	1/3
7	A-	B-	C+	1(--)	0	0	1	1/3
8	A-	B-	C-	1(--)	1(--)	1(--)	3	1

Now Bell's theorem asks: *What is the probability that the polarization at any neighbor will be the same as the first?* We also calculate the sum and average of the polarization. Assuming realism is true, then by looking at Table 4-1, the answer to the question must be the probability must be >= 1/3. This is what the Bell inequality gives: A means to put this assertion to the test. Here is the incredible part: Believe it or not quantum mechanics violates Bell's inequality giving probabilities less than 1/3. This was proven experimentally for the first time in 1982 by French physicist Alain Aspect.

So now let's translate the photon polarization above into an experiment that can be run in a quantum computer. In 1969 John Clauser, Michael Horne, Abner Shimony, and Richard Holt came up with a proof for Bell's theorem: The CHSH inequality which formally states:

$$S = \langle A, B \rangle - \langle A, B' \rangle + \langle A', B \rangle + \langle A', B' \rangle$$

$$S \leq 2$$

To illustrate this we have two detectors: Alice and Bob. Given A and A' are detector settings on side Alice, B and B' on side Bob, with the four combinations being tested in separate experiments. Realism says that for a pair of entangled particles the parity table showing all possible permutations looks as shown here:

A	B	1
A	B'	0
A'	B	0
A'	B'	1

In classical realism, the CHSH inequality becomes $|S| = 2$. However, the mathematical formalism of quantum mechanics predicts a maximum value for S of $|S|=2\sqrt{2}$, thus violating this inequality. This can be put to the test using four separate quantum circuits (1 per measurement) with 2 qubits each. To simplify things, let measurements on Alice's detector be A = Z, A' = X, and Bob's detector B = W, B' = V (see Table 4-2). To begin the experiment, a basis Bell state must be constructed which matches the identity (see Figure 4-5):

$$1/\sqrt{2}\left(|00\rangle + |11\rangle\right)$$

The preceding expression essentially means: The qubit held by Alice can be 0 or 1. If Alice measured her qubit in the standard basis, the outcome would be perfectly random, either possibility having probability 1/2. But if Bob then measured his qubit, the outcome would be the same as the one Alice got. So, if Bob measured, he would also get a random outcome on first sight, but if Alice and Bob communicated they would find out that, although the outcomes seemed random, they are correlated.

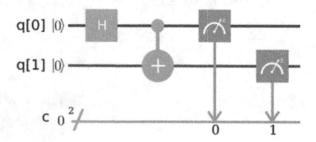

Figure 4-3. Basis Bell state

In Figure 4-3 two qubits are prepared in the ground state |0>. The H gate creates a superposition of the first qubit to the state $1/\sqrt{2}(|00\rangle + |10\rangle)$. Next, the CNOT gate flips the second qubit if the first is excited, resulting in the state $1/\sqrt{2}(|00\rangle + |11\rangle)$. This is the initial entangled state required for the four measurements in Table 4-2 (All reprints courtesy of International Business Machines Corporation, © International Business Machines Corporation).

- To rotate the measurement basis to the ZW axis, use the sequence of gates S-H-T-H.

- To rotate the measurement basis to the ZV axis, use the sequence of gates S-H-T'-H.

- The XW and XV measurement is performed the same way as above and the X via a Hadamard gate before a standard measurement.

Tip   Before performing the experiment in the composer, make sure its topology (the number of qubits and target device) in the score is set to two over a simulator. Some topologies (like the 5 qubit in a real quantum device) do not support entanglement for qubits 0 and 1 giving errors at design. Note that the target device can be a real quantum processor or a simulator.

Table 4-4. *Quantum circuits for Bell states*

Bell state measurement	Result for 100 shots	

AB (ZW)

c[2]	Probability
11	0.39
10	0.06
00	0.46
01	0.09

AB' (ZV)

c[2]	Probability
11	0.49
10	0.07
00	0.36
01	0.08

A'B (XW)

c[2]	Probability
11	0.42
10	0.05
00	0.49
01	0.04

A'B' (XV)

c[2]	Probability
11	0.05
10	0.52
00	0.03
01	0.40

Now we need to construct a table with the results of each measurement from Table 4-4 plus the correlation probability between A and B <AB>. The sum of the probabilities for the parity of the entangled particles is given by:

$$\langle AB \rangle = P(1,1) + P(0,0) - P(1,0) - P(0,1)$$

Remember that the ultimate goal is to determine if S ≤ 2 or |S| = 2; thus, by compiling the results of all measurements, we obtain Table 4-5.

Table 4-5. *Compiled results from the Bell experiment*

	P(00)	P(11)	P(01)	P(10)	<AB>
AB (ZW)	0.46	0.39	0.09	0.06	0.68
AB' (ZV)	0.36	0.49	0.08	0.07	0.73
A'B (XW)	0.49	0.42	0.04	0.05	0.47
A'B'(XV)	0.03	0.05	0.4	0.52	-0.32

Add the absolute values of column <AB> and we obtain |S| = 2.2. These results violate the Bell inequality (as predicted by quantum mechanics).

Three Qubit Entanglement with GHZ States Tests

These are named after physicists Greenberger-Horne-Zeilinger who came up with a generalization test for N entangled qubits with the simplest being a 3 qubit GHZ state:

$$|GHZ\rangle = 1 / \sqrt{2} \left(|000\rangle - |111\rangle \right)$$

Note   The importance of the GHZ states is that they show that the entanglement of more than two particles conflicts with local realism not only for statistical (probabilistic) but also nonstatistical (deterministic) predictions.

In simple terms, GHZ states show a stronger violation of Bell's inequality. Let's see how with a simple puzzle: imagine three independent boxes each containing two variables X, Y. Each variable has two possible outcomes: 1, -1. The question is to find a set of values for X, Y that solves the following set of identities:

(1) XYY = 1
(2) YXY = 1
(3) YYX = 1
(4) XXX = -1

For the impatient out there, there is no solution for this. For example, replace Y = 1 in (1) (2) and (3) then multiply them, that is, (5) = (1) * (2) * (3). The set then becomes:

(1) X11 = 1
(2) 1X1 = 1
(3) 11X = 1
(4) XXX = -1
(5) Multiply (1) (2) (3) and we get XXX = 1

There is no solution because identity (4) XXX = -1 contradicts identity (5) XXX = 1. The scary part is that a GHZ state can indeed provide a solution to this problem. This seems impossible in the deterministic view of classical reality, but nothing is impossible in the world of quantum mechanics, just improbable.

Incredibly, GHZ tests can rule out the local reality description with certainty after a single run of the experiment, but first we must construct a GHZ basis state. Try building the circuit in Table 4-6 in the composer to see the probabilities match the left side chart.

Table 4-6. *Basis GHZ state*

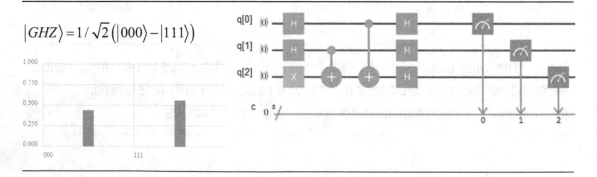

To kick start the experiment, the basis GHZ state (as well as probability results which should be around half) is shown in Table 4-6:

1. In the basis circuit, Hadamard gates in qubits 1 and 2 put them in superposition |00,01,01,11>. At the same time, the X gate negates qubit 3; thus, we end up with the states $1/\sqrt{2}(|001\rangle+|101\rangle+|011\rangle+|111\rangle)$.

2. The two CNOT gates entangle all qubits into the state $1/\sqrt{2}(|001\rangle+|010\rangle+|100\rangle+|111\rangle)$.

3. Finally, the three Hadamard gates map step 2 to the state $\frac{1}{2}(|000\rangle-|111\rangle)$.

Now, create the quantum circuits for identities XYY, YXY, XYY, and XXX from the previous section as shown in Table 4-7 (All reprints courtesy of International Business Machines Corporation, © International Business Machines Corporation).

Table 4-7. Quantum circuits for GHZ states

Measurement		Results for 100 shots	
YYX			
		c[3]	Probability
		011	0.34
		101	0.23
		110	0.23
		000	0.20
YXY			
		c[3]	Probability
		011	0.23
		101	0.28
		110	0.25
		000	0.24

(continued)

Table 4-7. (*continued*)

Measurement	Results for 100 shots

XYY

c[3]	Probability
011	0.23
101	0.26
110	0.35
000	0.16

XXX

c[3]	Probability
010	0.25
100	0.32
111	0.22
001	0.21

- For the measurement of X apply the H gate to the corresponding qubit.

- For each instance of Y apply the S† (S-dagger), and H gates to the corresponding qubit.

All in all, the principles of entanglement shown in this section were not popular in the early years of the 20th century. As a matter of fact, they were challenged by a theory called super determinism which sought to give a way out.

Super Determinism: A Way Out of the Spookiness. Was Einstein Right All Along?

In an interview for BBC in 1969, Physicist John Bell talked about his work on quantum mechanics. He said that we must accept the predictions that actions are transferred faster than the speed of light between entangled particles, but at the same time we cannot do anything with it. Information cannot travel faster than the speed of light, a fact that is also predicted by quantum mechanics. As if nature is playing a trick on us. He also mentioned that there is a way out of this riddle through a principle called super determinism.

Particle entanglement implies that measurements performed in one particle affect the other instantaneously, even across large distances (think opposite sides of the galaxy or the universe), and even across time. Einstein was an ardent opponent of this theory famously writing to Neils Bohr *God does not throw dice*. He could not accept the probabilistic nature of quantum mechanics so in 1935, along with colleagues Podolsky and Rosen, they came up with the infamous EPR paradox to challenge its foundation. In the EPR paradox if two entangled particles are separated by a tremendous distance, a measurement in one could not affect the other instantaneously as the event will have to travel faster than the speed of light (the ultimate speed limit in the universe). This will violate general relativity, thus creating a paradox: nothing travels faster than the speed of light that is the absolute rule of relativity.

Nevertheless, in 1982 the predictions of quantum mechanics were confirmed by French Physicist Alain Aspect. He devised an experiment that showed Bell's inequality is violated by entangled photons. He also proved that a measurement in one of the entangled photons travels faster than the speed of light to signal its state to the other. Since then, Aspect's results have been proven correct time and again (details on his experiment are shown in Chapter 1). The irony is that there is a chance that Einstein was right all along and entanglement is just an illusion. It is the principle of super determinism.

Tip   In simple terms super determinism says that freedom of choice has been removed since the beginning of the universe. All particle correlations and entanglements were established at the moment of the big bang. Thus, there is no need for a faster than light signal to tell particle B what the outcome of particle A is.

If true, this loophole will prove that Einstein was right when postulating the EPR paradox and all our hard work thus far is just an illusion. But this principle sounds more like religious dogma (all outcomes are determined by fate) than science as Bell argued that super determinism was implausible. His reason was that freedom of choice is effectively free for the purpose at hand due to alterations introduced by a large number of very small effects. Super determinism has been called untestable as experimenters would never be able to eliminate correlations that were created at the beginning of the universe. Nevertheless, this hasn't stopped scientists from trying to prove Einstein right and particle entanglement an illusion. As a matter of fact, there is an experiment hard at work to settle things up and is really inventive. Let's see how.

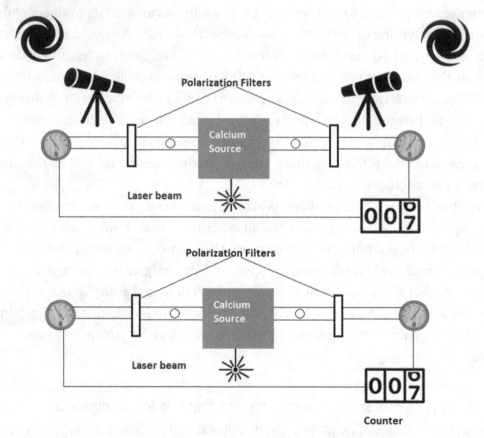

Figure 4-4. *Bell's inequality experiment using cosmic photons vs. the standard test*

Figure 4-4 shows the standard Bell inequality test experiment (at the bottom) and a variation of the experiment using cosmic photons (at the top) by Andrew Friedman and colleagues at MIT.[2]

---

[2] Jason Gallicchio, Andrew S. Friedman, and David I. Kaiser. Testing Bell's Inequality with Cosmic Photons: Closing the Setting-Independence Loophole. available online at https://arxiv.org/abs/1310.3288

Tip   For a full description of the standard Bell inequality test, see Chapter 1.

Friedman and colleagues came up with a novel variation of the standard Bell experiment using cosmic rays. The idea is to use real-time astronomical observations of distant stars in our galaxy, distant quasars, or patches of the cosmic microwave background, to essentially let the universe decide how to set up the experiment instead of using a standard quantum random number generator. That is, photons from distant galaxies are used to control the orientation of the polarization filters just before the arrival of entangled photons.

If successful, the implications would be groundbreaking. If the results from such an experiment do not violate Bell's inequality, it would mean that super determinism could be true after all. Particle entanglement will be an illusion, and signal transfer between entangled particles could not travel faster than light as predicted by relativity. Einstein will be right and there is no spooky action at a distance.

Luckily for quantum mechanics, no such thing has happened so far. Keep in mind that Friedman and colleagues are not the only team getting into the action. There are multiple teams trying to crack this riddle. As a matter of fact most of their results agree with quantum mechanics. That is, their results violate Bell's inequality. So it seems that the rift created by Einstein and Bohr in their struggle between relativity and quantum mechanics has not been settled yet. The next section shows how IBM Quantum can be accessed remotely via its slick REST API.

Remote Access via the REST API

Quantum features a relatively unknown REST API that handles all remote communications behind the scenes. It is used by the current Python SDKs:

- Qiskit: The Quantum Information Software Kit is the de facto access tool for quantum programming in Python.

- QiskitIBMQProvider: A lesser known library bundled with Qiskit that wraps the REST API in a Python client.

In this section, we peek inside IBMQprovider and look at the different REST endpoints for remote access. But first, authentication is required.

Authentication

To invoke any REST API call, we must first obtain an access token. This will be the access key to invoke any of the calls in this section. There are two ways of obtaining this token:

- Using your API token: To obtain the API token, login to the IBM Q console and copy it from the dashboard (see Figure 4-5).

- Using your account user name and password: Let's see how this is done using REST.

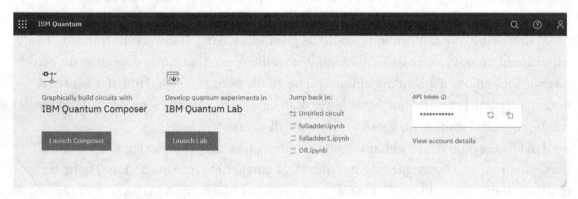

Figure 4-5. *Obtain your API token from the console.*

Authentication via API Token

- **HTTP method:** POST

- **URL:** https://auth.quantum-computing.ibm.com/api/users/ loginWithToken

- **Payload:** {"apiToken": "YOUR_API_TOKEN"}

Authentication via User-Password

- **HTTP method:** POST

- **URL:** https://auth.quantum-computing.ibm.com/api/users/login

- **Payload:** {"email": "USER-NAME", "password": "YOUR-PASSWORD"}

The response for both methods is:

```
{
  "id": "ACCESS_TOKEN",
  "ttl": 1209600,
  "created": "2018-04-15T20:21:03.204Z",
  "userId": "USER-ID"
}
```

where *id* is your access token, *ttl* is the time to live (or expiration time) in milliseconds, and *userId* is your user id. Save the access token and the user id for use in this section. Note that when your session expires, a new access token needs to be generated.

List Available Backends

This call returns a JSON list of all available backends and simulators in IBM Q:

- **HTTP method:** GET

- **URL:** `https://api-qcon.quantum-computing.ibm.com/api/Backends?access_token=ACESS-TOKEN`

Request Parameters

Name	Value
access_token	Your account access token

HTTP Headers

Name	Value
x-qx-client-application	Defaults to qiskit-api-py

Response Sample

The response content type for all API calls is application/json. The next paragraph shows the partial result of a call to this endpoint. Note that this endpoint will return both real processors and simulators.

```
[{
        "name": "ibmqx2",
        "version": "1",
        "status": "on",
        "serialNumber": "Real5Qv2",
        "description": "5 transmon bowtie",
        "basisGates": "u1,u2,u3,cx,id",
        "onlineDate": "2017-01-10T12:00:00.000Z",
        "chipName": "Sparrow",
        "id": "28147a578bdc88ec8087af46ede526e1",
        "topologyId": "250e969c6b9e68aa2a045ffbceb3ac33",
        "url": "https://ibm.biz/qiskit-ibmqx2",
        "simulator": false,
        "nQubits": 5,
        "couplingMap": [
                [0, 1],
                [0, 2],
                [1, 2],
                [3, 2],
                [3, 4],
                [4, 2]
        ]
},..]
```

The most important keys from the response above are described in Table 4-8.

Table 4-8. *Available backends response keys*

Key	Description
Name	The name id of the processor to be used when executing code against it
Version	A string or positive integer used to track changes to the processor
Description	This is probably a description of the hardware used to build the chip. You may see things like
	• 5 transmon bowtie
	• 16 transmon 2x8 ladder
	Note: A trasmon is defined as a type of noise-resistant superconducting charge qubit. It was developed by Robert J. Schoelkopf, Michel Devoret, Steven M. Girvin, and their colleagues at Yale University in 2007[3]
basisGates	These are the physical qubit gates of the processor. They are the foundation under which more complex logical gates can be constructed
nQubits	The number of qubits used by the processor
copulingMap	The coupling map defines interactions between individual qubits while retaining quantum coherence. It is used to simplify quantum circuitry and allow the system to be broken up into smaller units

Get Backend Parameters

This call returns a JSON list of the backend parameters for a given processor in Q Experience. Some of these parameters include

- Qubit cool down temperature in Kelvin degrees: For example, I got 0.021 K for ibmqx4 – that is a super frosty -459.6 F or -273.1 C.

- Buffer times in ns.

---

[3] J. Koch et al., "Charge-insensitive qubit design derived from the Cooper pair box," Phys. Rev. A 76, 04319 (2007), doi:10.1103/PhysRevA.76.042319, arXiv:0703002

- Gate times in ns.

- Other quantum specs are documented in more detail at the backend information site.[4]

The request type and endpoint URL are

- **HTTP method:** GET

- **URL:** https://api-qcon.quantum-computing.ibm.com/api/ Backends/NAME/properties?access_token=ACCESS-TOKEN

Request Parameters

Name	Value
access_token	Your account access token

HTTP Headers

Name	Value
x-qx-client-application	Defaults to qiskit-api-py

Response Sample

Listing 4-2 shows a simplified response for ibmqx4 parameters in JSON.

Listing 4-2. Simplified response for ibmqx4 parameters

```
{
    "lastUpdateDate": "2018-04-15T10:47:03.000Z",
    "fridgeParameters": {
        "cooldownDate": "2017-09-07",
        "Temperature": {
            "date": "2018-04-15T10:47:03Z",
```

---

[4] IBM Quantum backend information available online at https://github.com/QISKit/ ibmqx-backend-information

```
                    "value": 0.021,
                    "unit": "K"
            }
    },
    "qubits": [{
            "name": "Q0",
            "buffer": {
                    "date": "2018-04-15T10:47:03Z",
                    "value": 10,
                    "unit": "ns"
            },
            "gateTime": {
                    "date": "2018-04-15T10:47:03Z",
                    "value": 50,
                    "unit": "ns"
            },
            "T2": {
                    "date": "2018-04-15T10:47:03Z",
                    "value": 16.5,
                    "unit": "µs"
            },
            "T1": {
                    "date": "2018-04-15T10:47:03Z",
                    "value": 45.2,
                    "unit": "µs"
            },
            "frequency": {
                    "date": "2018-04-15T10:47:03Z",
                    "value": 5.24208,
                    "unit": "GHz"
            }
    },...]
```

Get the Status of a Processor's Queue

This call returns the status of a specific quantum processor event queue.

- **HTTP method:** GET

- **URL:** `https://api-qcon.quantum-computing.ibm.com/api/Backends/NAME/queue/status?access_token=ACCESS-TOKEN`

Request Parameters

It seems strange but this API call appears not to ask for an access token.

HTTP Headers

Name	Value
x-qx-client-application	Defaults to qiskit-api-py

Response Sample

For example, to get the event queue for ibmqx4, paste the following URL in your browser:
`https://quantumexperience.ng.bluemix.net/api/Backends/ibmqx4/queue/status`

The response looks like: `{"state":true,"status":"active","lengthQueue":0}` where

- **state:** It is the status of the processor. If alive true else false.

- **status:** It is the status of the execution queue: active or busy.

- **lengthQueue:** It is the size of the execution queue or the number of simulations waiting to be executed.

Tip   When you submit an experiment to IBM Q Experience, it will enter an execution queue. This API call is useful to monitor how busy the processor is at a given time.

List Jobs in the Execution Queue

This call returns a list of jobs in the processor execution queue.

- **HTTP method:** GET

- **URL:** https://api-qcon.quantum-computing.ibm.com/api/
  jobs?access_token=ACCESS-TOKEN&filter=FILTER

Request Parameters

Name	Value
access_token	Your account access token
filter	A result size hint in JSON. For example: {"limit":2} returns a maximum of 2 entries

HTTP Headers

Name	Value
x-qx-client-application	Defaults to qiskit-api-py

Response Sample

Listing 4-3 shows the response format for this call. The information appears to be a historical record of experiment executions containing information such as status, dates, results, code, calibration, and more.

Listing 4-3. Simplified response for the get jobs API call

```
[{
    "qasms": [{
        "qasm": "...",
        "status": "DONE",
        "executionId": "331f15a5eed1a4f72aa2fb4d96c75380",
        "result": {
            "date": "2018-04-05T14:25:37.948Z",
            "data": {
```

```
                    "creg_labels": "c[5]",
                    "additionalData": {
                            "seed": 348582688
                    },
                    "time": 0.0166247,
                    "counts": {
                            "11100": 754,
                            "01100": 270
                    }
                }
            }
        }
    }],
    "shots": 1024,
    "backend": {
        "name": "ibmqx_qasm_simulator"
    },
    "status": "COMPLETED",
    "maxCredits": 3,
    "usedCredits": 0,
    "creationDate": "2018-04-05T14:25:37.597Z",
    "deleted": false,
    "id": "d405c5829274d0ee49b190205796df87",
    "userId": "ef072577bd26831c59ddb212467821db",
    "calibration": {}
}, ...]
```

Note   Depending on the size of the execution queue, you may get an empty result ([]) if there are no jobs in the queue or a formal result as shown in Listing 4-3. Whatever the case, make sure the HTTP response code is 200 (OK).

Get Account Information

When an account is created, each user is assigned several execution credits which are spent when running experiments. This call lists your credit information.

- **HTTP method:** GET

- **URL:** https://auth.quantum-computing.ibm.com/api/users/USER-ID?access_token=ACCESS-TOKEN

Tip   The user id can be obtained from the authentication response via API token or user-password. See the "Authentication" section for details.

Request Parameters

Name	Value
access_token	Your account access token

HTTP Headers

Name	Value
x-qx-client-application	Defaults to qiskit-api-py

Response Sample

Listing 4-4 shows a sample response for this call.

Listing 4-4. Credit information sample response

```
{
    "institution": "Private Research",
    "status": "Registered",
    "blocked": "None",
    "dpl": {
```

```
        "blocked": false,
        "checked": false,
        "wordsFound": {},
        "results": {}
    },
    "credit": {
        "promotional": 0,
        "remaining": 150,
        "promotionalCodesUsed": [],
        "lastRefill": "2018-04-12T14:05:09.136Z",
        "maxUserType": 150
    },
    "additionalData": {
    },
    "creationDate": "2018-04-01T15:36:16.344Z",
    "username": "",
    "email": "",
    "emailVerified": true,
    "id": "",
    "userTypeId": "…",
    "firstName": "…",
    "lastName": "…"
}
```

List User's Experiments

This call lists all experiments for a given user id.

- **HTTP method:** GET

- **URL:** https://api-qcon.quantum-computing.ibm.com/api/users/ USER-ID/codes/lastest?access_token=ACCESS-TOKEN&includeExe cutions=true

Request Parameters

Name	Value
USER-ID	Your user id obtained from the authentication step
access_token	Your account access token
includeExecutions	If true, include executions in the result

HTTP Headers

Name	Value
x-qx-client-application	Defaults to qiskit-api-py

Response Sample

Listing 4-5 shows a sample response from this call.

Listing 4-5. Experiment list response

```
{
  "total": 17,
  "count": 17,
  "codes": [{
    "type": "Algorithm",
    "active": true,
    "versionId": 1,
    "idCode": "…",
    "name": "3Q GHZ State YXY-Measurement 1",
    "jsonQASM": {
      ...
    },
    "qasm": "",
    "codeType": "QASM2",
    "creationDate": "2018-04-14T19:09:51.382Z",
    "deleted": false,
```

```
    "orderDate": 1523733740504,
    "userDeleted": false,
    "displayUrls": {
        "png": "URL"
    },
    "isPublic": false,
    "id": "…",
    "userId": "…"
}]}
```

Run a Job on Hardware

Use this call to submit a Quantum Job to a hardware processor available in your customer plan.

- **HTTP method:** POST

- **URL:** `https://runtime-us-east.quantum-computing.ibm.com/jobs`

HTTP Headers

Name	Value
X-Access-Token	Access token
Content-Type	application/json

Payload Format

The format of the payload embeds all execution parameters: backend name, shots, and code in a single JSON document as shown in the following.

```
{
  "program_id": "circuit-runner",
  "hub": "ibm-q",
  "group": "open",
  "project": "main",
  "backend": "ibmq_quito",
```

```
"params": {
  "shots": 1024,
  "circuits": [
      " OPENQASM 2.0;\ninclude \"qelib1.inc\";\n\nqreg q[4];\ncreg c[4];\
      nh q[0];\nh q[1];\nx q[0];\nmeasure q[0] -> c[0];"
  ]
},
"tags": [
  "composer-info:composer:true",
  "composer-info:code-id:d2262dd4e07a9f894caabd977b10c7e98d90537228723dd1
  4dceaab023c36098",
  "composer-info:code-version-id:646d2a856260a179aa094d3e"
]
}
```

The preceding payload submits a random experiment to the real device ibm_quito. Make sure it is online before submission (or use the simulator instead). Also make sure the QASM code is in a single line including line feeds (\n). Note that double quotes must be escaped. If the submission fails, it probably means that the device is offline or your QASM payload is invalid.

Response Format

If everything goes OK, the response will look like:

```
{
    "id": "chn19jpike34bjj7d400",
    "backend": "ibmq_quito"
}
```

The id value represents the Job id. You can monitor its status in the Jobs menu of the dashboard (see Figure 4-6).

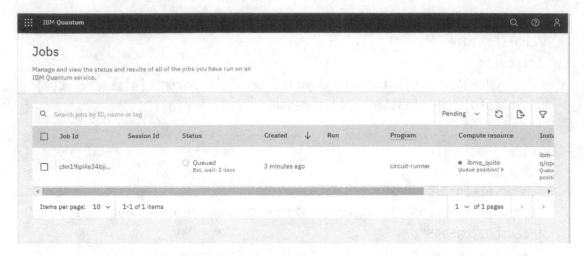

Figure 4-6. *Job status in the web console*

Note   Depending on the size of the execution queue, your job may sit for a while. Please keep the queue clean by canceling your test jobs.

Get the API Version

It returns the version of the Q Experience REST API.

- **HTTP method:** GET

- **URL:** https://api-qcon.quantum-computing.ibm.com/api/
  version?access_token=ACCESS-TOKEN

Request Parameters

Name	Value
access_token	Your account access token

HTTP Headers

Name	Value
x-qx-client-application	Defaults to qiskit-api-py

Response Format

It returns a string with the version of the API, by the time of this writing 0.153.0.

We have peaked inside the IBM Quantum REST API to see what goes on behind the scenes. Now let's put that knowledge to the test with a simple set of exercises.

Exercises

The IBM Quantum REST API can be a powerful tool to add functionality to your client apps. In this exercise set, we will use a platform desktop client to set up a REST workspace for the API calls we have seen in this chapter.

1. Install the Postman desktop client from `www.postman.com/`. Hint: When you start Postman, do not sign in to the cloud. This will allow you to work locally.

2. Create a new collection for the IBM Quantum REST API. Name the workspace *IBM Quantum*. **Tip: Read about the very useful Postman Global Variables to quickly store information such as API or access tokens.** See `https://learning.postman.com/docs/sending-requests/variables/`.

3. Create a new request: Authenticate via token. Use the following parameters:

   a. Type: POST

   b. URL: from the authentication section of this chapter: `https://auth.quantum-computing.ibm.com/api/users/loginWithToken`

c. Payload: {"apiToken": "YOUR_API_TOKEN"}. Tip: Copy the API token from the dashboard of the IBM Q console. Verify the Response returns an id (access token). **Please note that the API token (used for authentication) is not the same as the access token (used to invoke operations).** *Tip: Store the access token in a Postman global variable for use in further operations.*

4. Create a new request: Authenticate via password. Use the following parameters:

   a. Type: POST

   b. URL: from the authentication section of this chapter: `https://auth.quantum-computing.ibm.com/api/users/login`

   c. Payload: {"email": "USER-NAME", "password": "YOUR-PASSWORD"}. Verify the Response returns an id (access token).

5. Create a GET request to fetch the list of available backends:

   a. Type: GET

   b. URL: `https://api-qcon.quantum-computing.ibm.com/api/Backends?access_token=ACCESS_TOKEN`

   c. Verify the response information. Tip: Use the access token from Exercise 3. **Do not use the API token from the dashboard.**

6. Create a GET request to fetch processor parameters (use processor *ibm_perth* or your favorite name from the console):

   a. Type: GET

   b. URL: `https://api-qcon.quantum-computing.ibm.com/api/Backends/ibm_perth/properties?access_token=ACCESS_TOKEN`

   c. Verify the response JSON. Name your request Get Backend properties. Tip: Use a Global variable to store the processor name.

7. Create a GET request to fetch the queue status of the previous processor:

   a. URL: `https://api-qcon.quantum-computing.ibm.com/api/Backends/`*`ibm_perth`*`/queue/status?access_token=ACCESS_TOKEN`

   b. Type: GET

   c. Verify the JSON response.

8. Create a GET request to List Jobs in the execution queue:

   a. Name: List Jobs

   b. Type: GET

   c. URL: `https://api-qcon.quantum-computing.ibm.com/api/jobs?access_token=ACCESS_TOKEN`

   d. Verify the JSON response.

9. Create a GET request to List user experiments:

   a. Name: List Experiments

   b. Type: GET

   c. URL: `https://api-qcon.quantum-computing.ibm.com//api/users/USER_ID/codes/lastest?access_token=ACCESS_TOKEN` (Tip: The user id is returned in the authentication request).

   d. Verify the JSON response.

10. Create a GET request to get the API version:

   a. Type: GET

   b. Name: Get API Version

   c. URL: `https://api-qcon.quantum-computing.ibm.com/api/version?access_token=ACCESS_TOKEN`

   d. Verify the JSON response.

Finally, export your collection. This will be a helpful tool if you plan to integrate the IBM Quantum REST API into your existing desktop or web app. Your Postman workspace should look similar to Figure 4-7.

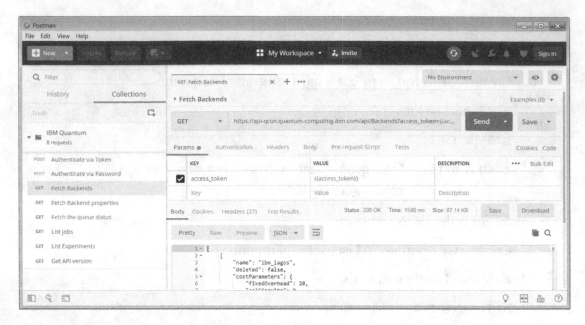

Figure 4-7. *Postman desktop app with the Quantum REST API collection*

CHAPTER 5

Mathematical Foundation: Time to Dust Up That Linear Algebra

Matrices, complex numbers, and tensor products are the holy trinity of quantum computing. During my quantum learning curve on the mathematical background, I wondered as you may: Where is the quantum magic in all this? How can a complex number or a matrix tell me about superposition or entanglement? Where is the spookiness? My immature understanding of superposition was that a system existed in multiple states at the same time: Like the cat in the box, it is dead and alive at once. However that is incorrect, the cat cannot be dead and alive at the same time just like a quantum system cannot exist in two simultaneous states.

My linear algebra was rusty to say the least. So after a while, I understood that superposition is simply an algebraic sum. I was a little disappointed after realizing this, and no, the cat is not dead and alive simultaneously: The cat is in a sum of probabilities: 50% death plus 50% alive. All in all, the bizarre properties of quantum mechanics: entanglement and superposition are completely described by matrices. It is the rich interpretation of matrices and complex numbers that allows for a bigger landscape, and it is what gives the quantum advantage over traditional scaler-based mathematics. Think of it as traveling from one city to another: scalers give a single road to travel. Matrices give countless paths, some of them bad, some good, and that is the key to the whole thing. That is what makes quantum algorithms superior to classical ones. Things get a little more complicated nonetheless, but at the end it's all linear algebra. So let's dust up that good old algebra.

© Vladimir Silva 2024
V. Silva, *Quantum Computing by Practice*, https://doi.org/10.1007/978-1-4842-9991-3_5

Qubit 101: Vector, Matrices, and Complex Numbers

Before we start digging into quantum programs and algorithms, we need to refresh some fundamental mathematics to understand what goes on behind the scenes. A fundamental way of understanding the basic model of the qubit and the effects of quantum gates is to use its algebraic representation. For this purpose you need some basic linear algebra concepts including

- Linear vectors: Simple vectors such as $\begin{bmatrix} 1 \\ 0 \end{bmatrix}$ which will be used to represent the basis states of the qubit.

- Complex numbers: A complex number is a number composed of real and imaginary parts denoted by a + bi where $i = \sqrt{-1}$. Note that the coefficients α, β of the super imposed state of a qubit $\psi = \alpha|0\rangle + \beta|1\rangle$ are complex numbers.

- Complex conjugate: A term that you will often hear when talking about quantum gates. To obtain a complex conjugate, simply flip the sign of the imaginary part; thus, a + bi becomes a − bi and vice versa.

- Matrix multiplication: if A is an n × m matrix and B is an m × p matrix, their product AB is an n × p matrix, in which the m entries across a row of A are multiplied with the m entries down a column of B and summed to produce an entry of AB. Take the first row from the first matrix and multiply each element for the first column of the second matrix, which becomes the first element in the result matrix.

- Matrix determinant det(M): It is a single numerical value useful when calculating the inverse or when solving a system of linear equations. For a 2x2 matrix, multiply the elements in the main diagonal, then subtract the product of the elements in the anti-diagonal: $A = \begin{bmatrix} a & b \\ c & d \end{bmatrix}$, $\det(A) = ad - bc$. Similarly, for 3x3 matrix, split in 3 2x2 determinants. Thus, given $A = \begin{bmatrix} a & b & c \\ d & e & f \\ g & h & i \end{bmatrix}$, $\det(A) = a\det\begin{bmatrix} e & f \\ h & i \end{bmatrix} - b\det\begin{bmatrix} d & f \\ g & i \end{bmatrix} + c\det\begin{bmatrix} d & e \\ g & h \end{bmatrix}$.

Note that there is a method to calculate the determinant for an NxN matrix but is complicated and outside the scope of this chapter.

Let's practice these concepts with a set of easy exercises.

Exercise 5.1: Find the product C of matrices $A = \begin{bmatrix} 1 & 2 & 3 \\ 4 & 5 & 6 \end{bmatrix}, B = \begin{bmatrix} a & b \\ c & d \\ e & f \end{bmatrix}$

$$AB = \begin{bmatrix} a+2c+3e & b+2d+3f \\ 4a+5c+6e & 4b+5d+6f \end{bmatrix}$$

Exercise 5.2: Remember that matrix product is not commutative: $AB \neq BA$. Show that this is the case for the previous matrices A, B.

$$BA = \begin{bmatrix} a & b \\ c & d \\ e & f \end{bmatrix} \begin{bmatrix} 1 & 2 & 3 \\ 4 & 5 & 6 \end{bmatrix} = \begin{bmatrix} a+4b & 2a+5b & 3a+6b \\ c+4d & 2c+5d & 3c+6d \\ e+4f & 2e+5f & 3e+ef \end{bmatrix}$$

Exercise 5.3: Calculate the determinant of $A = \begin{bmatrix} 1 & 2 \\ 3 & 4 \end{bmatrix}$

$$|A| = 1*4 - 3*2 = -2$$

Exercise 5.3a: Trickier – Calculate the determinant of the 3x3 matrix: $M = \begin{bmatrix} u & b & c \\ d & e & f \\ g & h & i \end{bmatrix}$

$$|M| = a\begin{bmatrix} e & f \\ h & i \end{bmatrix} - b\begin{bmatrix} d & f \\ g & i \end{bmatrix} + c\begin{bmatrix} d & e \\ g & h \end{bmatrix} = a(ei - hf) - b(di - gf) + c(dh - go)$$

Try these on your own. If you get stuck, answers are provided in the appendix.

Exercise 5.4: Represent the values: 1.2, 3.0, −0.1 as a column vector.

Inner product: The inner product (a generalization of the dot product) of two vectors a and b, represented as (a.b) or <a,b>, is a way to multiply them together, with the result being a scalar. Thus, $<(x1, x2,..)<(y1, y2,..)> = x1y1 + x2+y2+...$

Exercise 5.5: Find the inner product of the vectors: $a = \begin{bmatrix} 1 \\ 2 \\ 3 \end{bmatrix}, b = \begin{bmatrix} 4 \\ 5 \\ 6 \end{bmatrix}$.

Transpose of a Matrix M^T

The transpose of a matrix is a flipped version of the original matrix obtained by switching its rows with its columns.

Exercise 5.6: Find the transpose of $A = \begin{bmatrix} 1 & 2 & 3 \\ 4 & 5 & 6 \end{bmatrix}$

$$A^T = \begin{bmatrix} 1 & 4 \\ 2 & 5 \\ 3 & 6 \end{bmatrix}$$

Exercise 5.7: Try this one on your own. Find the transpose of $A = \begin{bmatrix} 1 \\ 2 \\ 3 \end{bmatrix}$.

Conjugate Transpose or Adjoint M†

The conjugate transpose, or Hermitian transpose, of an m × n complex matrix A is an n × m matrix obtained by transposing A and applying the complex conjugate on each entry. Where the complex conjugate of z = a + ib, is z* = a – ib (more about complex numbers in the next section).

Exercise 5.8: What is the conjugate transpose of $A = \begin{bmatrix} a+i \\ 2b-i \\ 3c \end{bmatrix}$

$$A^\dagger = \begin{bmatrix} a-i & 2b+i & 3c \end{bmatrix}$$

Try these on your own.

Exercise 5.9: What is the conjugate transpose vector $v = \begin{bmatrix} a \\ b \\ c \end{bmatrix}$, where a, b, and c are complex numbers?

Exercise 5.10: Given the complex vector from the previous question, what is $v^\dagger v$.

Exercise 5.11 Calculate the product of $\begin{bmatrix} a \\ b \\ c \end{bmatrix} \begin{bmatrix} x & y & z \end{bmatrix}$.

Complex Numbers: The Mathematical Magic Hats

You probably heard of the famous imaginary number $i = \sqrt{-1}$ or $i^2 = -1$. It was coined in the 17th century by French philosopher René Descartes to find a solution to a paradox such as this:

$$x^2 = -1$$

We know that any number squared must be positive so how can you solve the previous equation? Simply let $i = \sqrt{-1}$ then you have x = i. This is the equivalent of pulling a rabbit from the magician's hat. In fact, imaginary numbers are a fundamental part of the mathematical arsenal and an integral part of quantum mechanics. Complex numbers can be represented geometrically in the complex plane using a Cartesian coordinate system (see Figure 5-1).

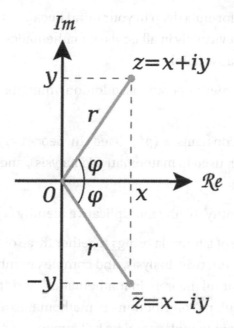

Figure 5-1. *Complex plane for the complex number Z, where the X-axis is the real part and the y-axis is the imaginary part*

Tip   We know that $x = r \cdot \cos(\varphi)$ and $y = r \cdot \sin(\varphi)$; therefore, we can write a complex number as $z = r \cos(\varphi) + i(r)\sin(\varphi)$. After factorization and using the almighty Euler identity formula (see next section), we obtain $z = r\, e^{\varphi}$. This is known as the ***polar form*** of a complex number where r is the magnitude and φ is the rotation angle in the complex plane. Try to remember this polar form and the superb Euler identity. They are extremely important.

Exercise 5.12: Here is a trick question: Which of the following is a square root of −16? a) -4, b) 4i.

Exercise 5.13: Find the powers of i: i^{-3}, i^{-2}, i^{-1}, i^{0}, i^{1}, i^{2}, i^{3}.

Euler's Identity: A Wonderful Masterpiece

A cornerstone of quantum computing is Euler's identity, developed by the Swiss mathematician of the same name. Here it is in all its glory:

$$e^{ix} = \cos x + i \sin x$$

You must engrave this formula deep in your mind because it is everywhere, not only in quantum mechanics but virtually in all fields of mathematics. What makes Euler's identity so incredible? Because it links

- Three basic arithmetic operations: addition, multiplication, and exponentiation

- Three powerful constants: π (pi, used in Geometry), e (Euler's number, used in mathematical analysis), and i (the imaginary number)

- The additive identity (0), the multiplicative identity (1)

Thus, this powerhouse of a formula brings together three of the biggest fields in mathematics: geometry, numerical analysis, and complex numbers. This is why it has been called "the most beautiful theorem in mathematics" and "the greatest equation ever" just to name a few. Euler's contributions to mathematics are too long to count, so let's dive inside this little gem to understand its brilliance.

Exercise 5.14: Euler's identity is closely related to the infinite series for the sine, cosine, and exponent where:

$$\sin x = x - \frac{x^3}{3!} + \frac{x^5}{5!} - \frac{x^7}{7!} + \dots (1)$$

$$\cos x = 1 - \frac{x^2}{2!} + \frac{x^4}{4!} - \frac{x^6}{6!} + \dots (2)$$

$$e^x = 1 + \frac{x^1}{1!} + \frac{x^2}{2!} + \frac{x^3}{3!} + \frac{x^4}{4!} + \frac{x^5}{5!} + \dots (3)$$

Prove Euler's identity using the preceding infinite series. Hint: Start with e^x and apply the powers of i: $i^{-3} = i$, $i^{-2} = -1$, $i^{-1} = -i$, $i^0 = 1$, etc. Finally group the terms that contain i and compare with the series for the cosine and sine.

$$e^{ix} = 1 + ix + \frac{(ix)^2}{2!} + \frac{(ix)^3}{3!} + \frac{(ix)^4}{4!} + \frac{(ix)^5}{5!} + \ldots$$

$$= \left[1 - \frac{x^2}{2!} + \frac{x^4}{4!} - \frac{x^6}{6!} + \ldots \right] + i \left[x - \frac{x^3}{3!} + \frac{x^5}{5!} - \frac{x^7}{7!} + \ldots \right] = \cos x + i \sin x$$

Try these on your own. Use the identities $\sin x = \sin(-x)$, $\cos x = \cos(-x)$, and Euler's. Hint: Apply Euler's to all exponentials, then simplify.

Exercise 5.15: Show that $\cos x = 1/2(e^{ix} + e^{-ix})$

Exercise 5.16: Show that $\sin x = 1/2i(e^{ix} - e^{-ix})$

All in all, Euler's identity is critical to our understanding of quantum computing. As a matter of fact, any single qubit gate can be generalized as a rotation over 3 Euler angles, truly a masterpiece.

$$U(\theta, \varphi, \lambda) = \begin{bmatrix} \cos\dfrac{\theta}{2} & -e^{i\lambda} \sin\dfrac{\theta}{2} \\ e^{i\varphi} \sin\dfrac{\theta}{2} & e^{i(\varphi+\lambda)} \cos\dfrac{\theta}{2} \end{bmatrix}$$

Tensor Product of a Matrix $\otimes$

Tensor products are important in areas of abstract algebra, algebraic topology, geometry, as well as differential geometry and physics. In quantum computing, pretty much everything is a tensor product of vectors or matrices (operators). You must get familiar with this concept. Tensor products are important because

1. They allow you to change the ring over which a module is defined.

2. They allow you to transform bilinear maps into linear ones, to which you can apply linear algebra.

In simple terms the tensor product of two matrices $A = \begin{bmatrix} a & b \\ c & d \end{bmatrix}, B = \begin{bmatrix} e & f \\ g & h \end{bmatrix}$ is

denoted by $A \otimes B = \begin{bmatrix} a\begin{vmatrix} e & f \\ g & h \end{vmatrix} & b\begin{vmatrix} e & f \\ g & h \end{vmatrix} \\ c\begin{vmatrix} e & f \\ g & h \end{vmatrix} & d\begin{vmatrix} e & f \\ g & h \end{vmatrix} \end{bmatrix}.$

Tip   In quantum mechanics, the tensor product is mostly abbreviated; thus, we can write |0>⊗|1> = |0>|1> = |01>.

Let's illustrate points 1 and 2 earlierwith a simple example using quantum gates.

Exercise 5.17: The Control-X Gate (CX) is made of the tensor product of the identity $CX = |0\rangle\langle 0| \otimes I + |1\rangle\langle 1| \otimes X$ where:

$I = \begin{bmatrix} 1 & 0 \\ 0 & 1 \end{bmatrix}, X = \begin{bmatrix} 0 & 1 \\ 1 & 0 \end{bmatrix}, |0\rangle = \begin{bmatrix} 1 \\ 0 \end{bmatrix}, |0\rangle = \begin{bmatrix} 1 & 0 \end{bmatrix}, |1\rangle = \begin{bmatrix} 0 \\ 1 \end{bmatrix}, \langle 1| = \begin{bmatrix} 0 & 1 \end{bmatrix}$. Calculate the tensor product

of the identity and verify the result with the matrix representation for $CX = \begin{bmatrix} 1 & 0 & 0 & 0 \\ 0 & 1 & 0 & 0 \\ 0 & 0 & 0 & 1 \\ 0 & 0 & 1 & 0 \end{bmatrix}.$

$$|0\rangle\langle 0| = \begin{bmatrix} 1 \\ 0 \end{bmatrix}\begin{bmatrix} 1 & 0 \end{bmatrix} = \begin{bmatrix} 1 & 0 \\ 0 & 0 \end{bmatrix}, |1\rangle\langle 1| = \begin{bmatrix} 0 \\ 1 \end{bmatrix}\begin{bmatrix} 0 & 1 \end{bmatrix} = \begin{bmatrix} 0 & 0 \\ 0 & 1 \end{bmatrix}$$

$$CX = \begin{bmatrix} 1 & 0 \\ 0 & 0 \end{bmatrix}\otimes\begin{bmatrix} 1 & 0 \\ 0 & 1 \end{bmatrix}+\begin{bmatrix} 0 & 0 \\ 0 & 1 \end{bmatrix}\otimes\begin{bmatrix} 0 & 1 \\ 1 & 0 \end{bmatrix} = \begin{bmatrix} 1\begin{bmatrix} 1 & 0 \\ 0 & 1 \end{bmatrix} & 0\begin{bmatrix} 1 & 0 \\ 0 & 1 \end{bmatrix} \\ 0\begin{bmatrix} 1 & 0 \\ 0 & 1 \end{bmatrix} & 0\begin{bmatrix} 1 & 0 \\ 0 & 1 \end{bmatrix} \end{bmatrix}+\begin{bmatrix} 0\begin{bmatrix} 0 & 1 \\ 1 & 0 \end{bmatrix} & 0\begin{bmatrix} 0 & 1 \\ 1 & 0 \end{bmatrix} \\ 0\begin{bmatrix} 0 & 1 \\ 1 & 0 \end{bmatrix} & 1\begin{bmatrix} 0 & 1 \\ 1 & 0 \end{bmatrix} \end{bmatrix}$$

$$CX = \begin{bmatrix} 1 & 0 & 0 & 0 \\ 0 & 1 & 0 & 0 \\ 0 & 0 & 0 & 1 \\ 0 & 0 & 1 & 0 \end{bmatrix}$$

Postulates of Quantum Mechanics

This section may be a little tough for the not mathematically inclined; however, we need to dive a little, not too deep into quantum mechanics to have a solid understanding and complement the linear algebra.

Postulate 1: State and Vector Space

A quantum system is completely described by a state vector in Hilbert space:

$$\left|\Psi\right\rangle = \sum_i \alpha_i \left|e_i\right\rangle, \sum_i \left|\alpha^2\right| = 1$$

This postulate is telling us that any quantum state ψ is described by a linearly independent superposition (sum) of states e_i multiplied by a complex coefficient α_i. A very important requirement is that the sum of the squares of the complex coefficients must be 1. In quantum computing all transformations must be unitary (reversible) and this property ensures that. Another important concept to remember is the *basis vector*. For a single qubit, the basis vectors are $\left|0\right\rangle = \begin{bmatrix} 1 \\ 0 \end{bmatrix}, \left|1\right\rangle = \begin{bmatrix} 0 \\ 1 \end{bmatrix}$. For a 2 qubit system we have $2^2 = 4$ basis vectors:

$$\left|00\right\rangle = \begin{bmatrix} 1 \\ 0 \\ 0 \\ 0 \end{bmatrix}, \left|01\right\rangle = \begin{bmatrix} 0 \\ 1 \\ 0 \\ 0 \end{bmatrix}, \left|10\right\rangle = \begin{bmatrix} 0 \\ 0 \\ 1 \\ 0 \end{bmatrix}, \left|11\right\rangle = \begin{bmatrix} 0 \\ 0 \\ 0 \\ 1 \end{bmatrix}$$

Finally, an n-quibit system will have 2^n basis vectors. Note that they are linearly independent (only 1 element in the column contains a 1).

Postulate 2: Observables and Operators

Every observable of a physical system is described by an operator that acts on the states that describe the system.

- Observable: Is a measurable property of a particle such as spin, position, polarization, or charge.

- Operator: Is a synonym for Hermitian matrix which is a complex square matrix equal to its conjugate transpose.

Operators give rise to the concept of eigenstate and eigenvalue.

- Eigenstate: It is a quantum state changed only by a scalar multiplier. Thus for operator M acting on state |a>: $M \, | \, a \rangle = \lambda \, | \, a \rangle$.

- Eigenvalue: It is the scalar λ that corresponds to the eigenstate |a>.

Postulate 3: Measurement

When an observable is measured, the only possible outcome will be an eigenvalue of the operator for that observable.

Note   Physical measurements are always real numbers; any imaginary part will vanish.

- Measurement of eigenstates: The eigenstates a_j of an operator are orthogonal (perpendicular) and form a basis. $\langle a_j | a_k \rangle = \delta_{j,k}$. Any state can be expressed as a linear combination with complex coefficients of these eigenstates.

- Probability amplitude: When a system |ψ> is measured, the probability of obtaining eigenvalue λ for eigenstate |a> is: $|\langle a_j | \psi \rangle|^2$. Note that <a|ψ> is a complex value (or probability amplitude). It must be squared to obtain a real value.

- Expected value: It is the estimated result of running a measurement multiple times on state |ψ> using operator A: $\langle \Psi \rangle \equiv \sum_i \lambda_i \, \Pr(|a_j\rangle) = \langle \Psi | A | \Psi \rangle$

Postulate 4: Collapse of the Wave Function

The wave function for state |ψ> "collapses" on measurement and all information embodied in the superposition is lost permanently.

Note   Nobody knows why or how the wave function collapses. This is one of the great mysteries in physics. One bizarre theory, for example, proposes it is the result of energy transfer among parallel universes.

The collapse of the wave function implies a very ominous truth (depending on the point of view at least): We cannot "peek" inside the box, because doing so will destroy its quantum state. This is bad news for debugging a quantum circuit but good news for cryptographers where maintaining privacy is important.

Postulate 5: Unitary Transformations

A unitary transformation describes the evolution of a closed quantum system over time. Thus, given the state |ψ> at time t1 and the state |ψ2> at state t2 with unitary U depending only on t1, t2: $|\psi2\rangle = U \mid \psi\rangle$.

Tip   Unitary transformations ensure that the sum of probabilities for all states equals 1. All operations are reversible (except measurement), and the system is closed (it has no interactions with the environment).

Linear Algebra and Quantum Mechanics Cheat Sheet

Table 5-1 is a handy compilation of the most important concepts we have seen thus far. Keep it handy as you work through the exercises at the end of the chapter.

Table 5-1. *Linear algebra and quantum mechanics cheat sheet*

Name	Description
Bra-ket vector notation Bra = Row vector <a| Ket = column vector |a> Bra is the adjoint (complex conjugate transpose) of a ket	$\langle 0\|=\begin{bmatrix}1 & 0\end{bmatrix}\langle 1\|=\begin{bmatrix}0 & 1\end{bmatrix}$ $\|0\rangle=\begin{bmatrix}1\\0\end{bmatrix}\|1\rangle=\begin{bmatrix}0\\1\end{bmatrix}$
Basis vectors and standard basis are linearly independent; they cannot be written as a linear combination of other vectors Any vector can be written as a linear combination of basis vectors: |ψ> = c1{00> + c2|01> + c3|10> + c4|11>	Two qubit basis vectors: $\|00\rangle=\begin{bmatrix}1\\0\\0\\0\end{bmatrix}\|10\rangle=\begin{bmatrix}0\\1\\0\\0\end{bmatrix}\|01\rangle=\begin{bmatrix}0\\0\\1\\0\end{bmatrix}\|11\rangle=\begin{bmatrix}0\\0\\0\\1\end{bmatrix}$
Inner product <a|b> results in a scalar	$\langle 0\|0\rangle=\begin{bmatrix}1 & 0\end{bmatrix}\begin{bmatrix}1\\0\end{bmatrix}=1$
Outer product |a><b|: Can be thought as an operator that transforms |b> to |a>	$\|0\rangle\langle 0\|=\begin{bmatrix}1 & 0\\0 & 0\end{bmatrix}\|1\rangle\langle 1\|=\begin{bmatrix}0 & 0\\0 & 1\end{bmatrix}$ $\|1\rangle\langle 0\|=\begin{bmatrix}0 & 1\\0 & 0\end{bmatrix}\|0\rangle\langle 1\|=\begin{bmatrix}0 & 0\\1 & 0\end{bmatrix}$
Unitary matrix: It is a square complex matrix whose adjoint equals its inverse	$U=\begin{bmatrix}a & b\\c & d\end{bmatrix}U^{\dagger}=\begin{bmatrix}a^{*} & b^{*}\\c^{*} & da^{*}\end{bmatrix},\ UU^{\dagger}=U^{\dagger}U=I$ Hermitian: $U = U^{\dagger}$
State vector (wave function) |ψ> Superposition	$\|\psi\rangle=\sum_i \alpha_i \| e_i\rangle,\ \sum_i \|\alpha\|^2=1$ $\|\psi\rangle=\alpha\|0\rangle+\beta\|1\rangle$ *where* $\|\alpha\|^2+\|\beta\|^2=1$

(continued)

Table 5-1. (*continued*)

Name	Description
Tensor product of two operators A⊗B $$A = \begin{bmatrix} a & b \\ c & d \end{bmatrix}, B = \begin{bmatrix} e & f \\ g & h \end{bmatrix}$$ $$A \otimes B = \begin{bmatrix} a\begin{vmatrix} e & f \\ g & h \end{vmatrix} & b\begin{vmatrix} e & f \\ g & h \end{vmatrix} \\ c\begin{vmatrix} e & f \\ g & h \end{vmatrix} & d\begin{vmatrix} e & f \\ g & h \end{vmatrix} \end{bmatrix}$$	$$A \otimes B = \begin{bmatrix} ae & af & be & bf \\ ag & ah & bg & bh \\ ce & cf & de & df \\ cg & ch & dg & dh \end{bmatrix}$$
Entangled vs separable states: Entangled states cannot be expressed as the tensor product of independent qubit states	$\|\Psi\rangle = \frac{1}{\sqrt{2}}\|00\rangle + \frac{1}{\sqrt{2}}\|11\rangle\,(entangled)$ $\|\Psi\rangle = \frac{1}{\sqrt{2}}\|01\rangle + \frac{1}{\sqrt{2}}\|11\rangle\,(separable)$
Eigenstate and eigenvalue	$M\|a\rangle = \lambda\|a\rangle$ λ (lambda) is the eigenvalue for eigenstate la>
Pauli operators	$X\|0\rangle = \|1\rangle, X\|1\rangle = \|0\rangle$ $Y\|0\rangle = \|0\rangle, Y\|1\rangle = -\|i\rangle$ $Z\|0\rangle = \|0\rangle, Z\|1\rangle = -\|1\rangle$
Measurement probability for state la>> Note: <alψ> is the probability amplitude	$\|\langle a_i\| \psi\rangle\|^2$
Expected value <ψ> after repeated measurements is always different	$\langle\Psi\rangle \equiv \sum_i \lambda_i \, \mathrm{Pr}(\|a_j\rangle) = \langle\Psi\|A\|\Psi\rangle$
Unitary transformation • Preserves norm: the sum of probabilities = 1 • Ensures the system is reversible (required by quantum mechanics) • Ensures a closed system (no interaction with the environment)	$\|\psi2\rangle = U\|\psi\rangle$

A solid understanding of the algebra will help you realize how a quantum circuit can achieve massive parallelism and advantage over the traditional bit/transistor logic of the classical computer. Next, let's dive into basic gates and circuits.

Algebraic Representation of the Qubit

In the classical model, the fundamental unit of information is the bit which is represented by a zero or one. The bit physically translates to the voltage flow through a transistor. In quantum computation, the fundamental unit is the quantum bit (qubit) which physically translates to manipulations on photons, electrons, or atoms. Algebraically, the qubit is represented by Ket notation.

Tip   Ket notation was introduced in 1939 by Physicist Paul Dirac and is also known as the Dirac notation. The ket is typically represented as a column vector and written $|\varphi\rangle$.

Dirac's Ket Notation

Using Dirac's notation the basic quantum states of the qubit are represented by the vectors |0> and |1>. These are called the computational basis states.

Tip   The quantum state of a qubit is a vector in a two-dimensional complex vector space. Take a look at the following simple graph.

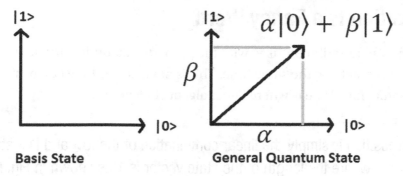

Figure 5-2. *Quantum states of the qubit*

Figure 5-2 shows the complex vector space used to represent the state of a qubit. On the left side, the so-called basis sate is made up of two unit vectors in Dirac notation for the states |0> and |1>. On the right side, a general quantum state is made up of a linear combination of the two. Thus, the basis states and general quantum states can be written as vectors:

$$|0\rangle = \begin{bmatrix} 1 \\ 0 \end{bmatrix}, |1\rangle = \begin{bmatrix} 0 \\ 1 \end{bmatrix}$$

$$\alpha |0\rangle + \beta |1\rangle$$

where α and β are amplitude coefficients of the unit vector. Note that a unit vector's amplitude must be 1; therefore, α and β must obey the constraint $|\alpha|^2 + |\beta|^2 = 1$. This algebraic representation is the key to understanding the effect of a logic gate in the qubit as you will see later on.

So why is the state of a qubit represented as a vector in a seemingly more complicated representation than its classical counterpart? Why vectors at all? The reason is that it allows for building a better model of computation as will be shown once we look at quantum gates and superposition of states. All in all, quantum mechanics is a theory that has evolved over many decades, and at the end of the day, a vector is a very simple mathematical object, easy to understand and manipulate. Probably the best tool for the job.

Superposition Is a Fancy Word

Superposition can be easily confused with the property of atomic particles (or qubits) to exist in multiple states at the same time. This is not true; nothing can be in multiple states at the same time. Here is where linear algebra can help.

Tip   Superposition is simply the linear combination of the I0> and I1> states. That is, $\alpha|0\rangle + \beta|1\rangle$ where the length of the state vector is 1 as shown in Figure 5-3.

Kets Are Column Vectors

If you find the ket notation confusing, just use the familiar vector representation instead. Thus, the superposition from the previous section can be written as:

$$|\Psi\rangle = \alpha|0\rangle + \beta|1\rangle = \alpha\begin{bmatrix}1\\0\end{bmatrix} + \beta\begin{bmatrix}0\\1\end{bmatrix} = \begin{bmatrix}\alpha\\\beta\end{bmatrix}$$

Note that, because kets are vectors, they obey the same rules as vectors do, for example, multiplication by a scalar:

$$2\left(\alpha|0\rangle + \beta|1\rangle\right) = 2\begin{bmatrix}\alpha\\\beta\end{bmatrix} = \begin{bmatrix}2\alpha\\2\beta\end{bmatrix}$$

Orient Yourself in the Bloch Sphere

Let's go beyond the algebraic 2D representation of the qubit; enter the Bloch sphere: a 3D representation of a single pure state (noiseless) qubit. Here is where the Euler formula you saw in the previous section shines. Consider Figure 5-3; let's define some important parts in the sphere.

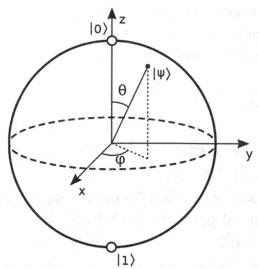

Z axis: known as Z-basis, or computational basis: |0>, |1>

X-axis: known as Fourier or Hadamard basis: |+>, |->.

Y-axis: known as imaginary basis: |i>, |-i>.

The state of the qubit |ψ> can be described by the rotation of two angles α[0,π] and φ[0,2π] where:

$$|\psi\rangle = cos\frac{\theta}{2}\ |0\rangle + e^{i\varphi}Sin\frac{\theta}{2}\ |1\rangle$$

Figure 5-3. *The Bloch sphere named after Swiss-American physicist Felix Bloch is a geometrical representation of a single pure state qubit*

Euler's magic formula unlocks the secret of describing |ψ> in the sphere. This is how:

- We start with the algebraic representation of $|\psi\rangle = \alpha|0\rangle + \beta|1\rangle$ where α, β are complex coefficients.

- Next, use Euler's formula for a complex in polar form $z = re^{i\varphi}$; thus, the state becomes:

$$|\Psi\rangle = r1e^{i\varphi1}|0\rangle + r2e^{i\varphi2}|1\rangle$$

- We can extract the term $e^{\varphi1}$ then we have:

$$|\Psi\rangle = e^{i\varphi1}\left[\ r1|0\rangle + r2e^{i(\varphi2-\varphi1)}|1\rangle\right]$$

- The term $ei^{\varphi1}$ is a global phase. Global phases cannot be measured and therefore can be discarded. Also we can rename $\varphi = \varphi2 - \varphi1$. Apply these rules to obtain:

$$|\Psi\rangle = r1|0\rangle + r2e^{i\varphi}|1\rangle$$

- Finally, the normalization condition: $r1^2 + r2^2 = 1$ implies that $r1 = \cos(\theta/2)$ and $r2 = \sin(\theta/2)$ where $\theta/2$ is an arbitrary angle selected for reasons that will become apparent soon. Thus, by replacing r1 and r2, we obtain:

$$|\Psi\rangle = \cos\frac{\theta}{2}\,|0\rangle + e^{i\varphi}\sin\frac{\theta}{2}|1\rangle$$

Tip   Euler's formula allows us to describe the pure state |ψ> of the qubit using only two polar angles α[0,π] and φ[0,2π] instead of the four parameters: r1, r2, and (θ, φ). Also note that, in the sphere, the basis vectors |0>, |1> point in opposite directions (anti-polar) whereas they are orthogonal (perpendicular) in the 2D Cartesian plane. This is why the angle in |ψ> is θ/2.

Exercise 5.18: Calculate the state |ψ> for the basis vectors of the computational basis |0>, |1>. Hint: Look at the sphere to see that |0>: θ = 0, φ = 0, |1>: θ = π, φ = 0.

$$|0\rangle \rightarrow \cos(0)|0\rangle + \sin(0)|1\rangle = |0\rangle$$

$$|1\rangle \rightarrow \cos(\pi/2)|0\rangle + \sin(\pi/2)|1\rangle = |1\rangle$$

Exercise 5.19: Calculate the state |ψ> for the basis vectors of the Hadamard basis |+>, |->. Hint: For |+>: θ = π/2, φ = 0, |->: θ = π/2, φ = π.

Exercise 5.20: Calculate the state |ψ> for the basis vectors of the imaginary basis |i>, -|i>. Hint: For |i>: θ = π/2, φ = π/2, -|i>: θ = π/2, φ = -π/2.

Changing the State of a Qubit with Quantum Gates

The purpose of quantum gates is to manipulate the state of a qubit to achieve a desired result. They are the basic building blocks of quantum computation just as classic logic gates are for the classical world. Some quantum gates are the equivalent of their classical counterparts. Let's take a look.

NOT Gate (Pauli X)

This is the simplest gate and it acts in a single qubit. It is the quantum equivalent of the classical NOT gate, and just like its counterpart, it flips the state of the qubit. Thus:

$$|0> \rightarrow |1>, |1> \rightarrow |0>$$

For a superposition, the X gate acts linearly, meaning it flips the corresponding state; thus, |0> becomes |1> and |1> becomes |0>; thus: $\alpha|0\rangle + \beta|1\rangle \rightarrow \alpha|1\rangle + \beta|0\rangle$

In a quantum circuit, the NOT gate is represented by an X or Pauli X, named after Austrian physicist Wolfgang Pauli, one of the fathers of quantum mechanics.

The circuit starts with the basis state |0> for qubit 0, the state flows through the quantum wire until a manipulation is done in the state, then the output continues through the wire.

There is another way of looking at the X gate in action, by using its matrix representation we can see exactly how the state is flipped by using the Pauli

Matrix $X = \begin{bmatrix} 0 & 1 \\ 1 & 0 \end{bmatrix}$.

The state of the qubit is flipped by using the matrix representation of X and the vectors for $|0\rangle = \begin{bmatrix} 1 \\ 0 \end{bmatrix}$ and $|1\rangle = \begin{bmatrix} 0 \\ 1 \end{bmatrix}$ thus:

$$X|0\rangle = \begin{bmatrix} 0 & 1 \\ 1 & 0 \end{bmatrix} \begin{bmatrix} 1 \\ 0 \end{bmatrix} = \begin{bmatrix} 0+0 \\ 1+0 \end{bmatrix} = \begin{bmatrix} 0 \\ 1 \end{bmatrix} = |1\rangle$$

$$X|1\rangle = \begin{bmatrix} 0 & 1 \\ 1 & 0 \end{bmatrix} \begin{bmatrix} 0 \\ 1 \end{bmatrix} = \begin{bmatrix} 0+1 \\ 0+0 \end{bmatrix} = \begin{bmatrix} 1 \\ 0 \end{bmatrix} = |0\rangle$$

There is an even simpler quantum circuit, the simplest of them all, and it is the quantum wire denoted by the Greek symbol (Psi) $|\psi\rangle$ _ _ _ _ _ _ _ _ _ $|\psi\rangle$ which describes the computational state over time. It may seem trivial, but physically this is the hardest thing to implement. Because of the atomic scale of the quantum wire (think photons, electrons, or single atoms), it is very fragile and prone to errors introduced by the environment.

Another interesting property of the X gate is that two NOT gates in a row give the identity matrix (I), a very important tool in linear transformations. Let's do the math:

q[0] |0⟩ —[Z]—[X]—

$|\psi\rangle \rightarrow XX|\psi\rangle$

To understand the effects of the circuit, let's see what happens when we multiply two X matrices:

$$XX = \begin{bmatrix} 0 & 1 \\ 1 & 0 \end{bmatrix}\begin{bmatrix} 0 & 1 \\ 1 & 0 \end{bmatrix} = \begin{bmatrix} 0+1 & 0+0 \\ 0+0 & 1+0 \end{bmatrix} = \begin{bmatrix} 1 & 0 \\ 0 & 1 \end{bmatrix} = I$$

The X gate is the simplest example of a quantum logic gate, circuit and computation. In the next section, we look at a truly quantum gate: Hadamard and how it can trigger super positions using circuits and algebra.

Truly Quantum: Super Positions with the Hadamard Gate

The effects of the Hadamard gate in the basis states are formally defined as:

$$|0\rangle \rightarrow \frac{|0\rangle + |1\rangle}{\sqrt{2}}, |1\rangle \rightarrow \frac{|0\rangle - |1\rangle}{\sqrt{2}}$$

Furthermore, for a superposition state $\alpha|0\rangle + \beta|1\rangle$ the Hadamard maps to:

$$\alpha|0\rangle + \beta|1\rangle \rightarrow \alpha\left(\frac{|0\rangle + |1\rangle}{\sqrt{2}}\right) + \beta\left(\frac{|0\rangle - |1\rangle}{\sqrt{2}}\right) = \frac{\alpha + \beta}{\sqrt{2}}|0\rangle + \frac{\alpha - \beta}{\sqrt{2}}|1\rangle$$

For the circuit and matrix presentation, the Hadamard acts on a single qubit.

q[0] |0⟩ —[H]—

Applying H to the basis states $|0\rangle = \begin{bmatrix} 1 \\ 0 \end{bmatrix}$ and $|1\rangle = \begin{bmatrix} 0 \\ 1 \end{bmatrix}$:

$$H = \frac{1}{\sqrt{2}}\begin{bmatrix} 1 & 1 \\ 1 & -1 \end{bmatrix}$$

$$H|0\rangle = \frac{1}{\sqrt{2}}\begin{bmatrix} 1 & 1 \\ 1 & -1 \end{bmatrix}\begin{bmatrix} 1 \\ 0 \end{bmatrix} = \frac{1}{\sqrt{2}}\begin{bmatrix} 1 \\ 1 \end{bmatrix} = \frac{1}{\sqrt{2}}\left(\begin{bmatrix} 1 \\ 0 \end{bmatrix} + \begin{bmatrix} 0 \\ 1 \end{bmatrix}\right) = \frac{|0\rangle + |1\rangle}{\sqrt{2}}$$

$$H|1\rangle = \frac{1}{\sqrt{2}}\begin{bmatrix} 1 & 1 \\ 1 & -1 \end{bmatrix}\begin{bmatrix} 0 \\ 1 \end{bmatrix} = \frac{1}{\sqrt{2}}\begin{bmatrix} 1 \\ -1 \end{bmatrix} = \frac{1}{\sqrt{2}}\left(\begin{bmatrix} 1 \\ 0 \end{bmatrix} - \begin{bmatrix} 0 \\ 1 \end{bmatrix}\right) = \frac{|0\rangle - |1\rangle}{\sqrt{2}}$$

So what is the computational reason for the Hadamard gate? What does this buy us? Without getting too technical, the answer is that the Hadamard gate expands the range of states that are possible for a quantum circuit. This is important because the expansion of states creates the possibility of finding shortcuts therefore doing computations faster. An analogy would be to a game of chess. For example, if your knight was allowed to move like a queen and knight at the same time (an expansion of states), this will tilt the game in your favor and allow you to checkmate faster. This is what Hadamard gives: more horsepower to our quantum machine.

Measurement of a Quantum State Is Trickier Than You Think

Imagine you have a lab in the basement of your home. You are given a qubit in state $|\psi\rangle = \alpha|0\rangle + \beta|1\rangle$, a measurement apparatus and asked to calculate the α and β coefficients. That is, compute the quantum state. It may seem like a trivial task; however, this is not possible. The principles of quantum mechanics state that the quantum state of a system is not directly observable. The best we can do is guess approximate information about α and β. This process is called measurement in the computational basis.

The outcome of a measurement on the quantum state $|\psi\rangle = \alpha|0\rangle + \beta|1\rangle$ gives the classical bits:

$\alpha|0\rangle + \beta|1\rangle \rightarrow 0$ *with probability* $|\alpha^2|$

$\alpha|0\rangle + \beta|1\rangle \rightarrow 1$ *with probability* $|\beta^2|$

Thus, the measurement process spits the probabilities of the classical bits 0 and 1 equal to the absolute values of the coefficients α and β squared. Physically, the way to imagine this process taking place is by observing a physical photon, atom, or electron with a measurement apparatus. This is the reason why measurement is often regarded as a quantum gate.

Measurement disturbs the state of the quantum system giving a classical bit outcome. The important thing to remember is that, after the process, the coefficients α and β are destroyed. This means that we cannot store large amounts of information in a qubit. Imagine if we could measure the exact values for α and β, then by using complex numbers, it would be possible in theory to store infinite amounts of classical information in the qubit state. By calculating the exact values of α and β, we could extract all that classical information. However, this is not possible. Quantum mechanics forbids it.

One final point on measurement is the *normalization of the quantum state*: Given a measurement in the computational basis $\alpha|0\rangle + \beta|1\rangle$, then the probability of the classical bit 0 and 1 must add to one. That is:

$$Probability\left(0\right) + Probability\left(1\right) = \left|\alpha^2\right| + \left|\beta^2\right| = 1$$

This means that the length of the quantum state vector must be one (normalized). This comes from the fact that measurement probabilities add to one. In the next section we'll talk about how single qubit gates are generalized, what they are, and how they are used to build more complex circuits.

Generalized Single Qubit Gates

So far we have seen two simple gates: X and H represented by the matrices:

$$X = \begin{bmatrix} 0 & 1 \\ 1 & 0 \end{bmatrix}, H = \frac{1}{\sqrt{2}} \begin{bmatrix} 1 & 1 \\ 1 & -1 \end{bmatrix}$$

Remember also that the superposition of the quantum state is expressed as a the vector $|\Psi\rangle = \begin{bmatrix} \alpha \\ \beta \end{bmatrix}$. Then applying both gates to the quantum state can be generalized for any unitary matrix:

$$H\begin{bmatrix} \alpha \\ \beta \end{bmatrix}, \; X\begin{bmatrix} \alpha \\ \beta \end{bmatrix}, \; U\begin{bmatrix} \alpha \\ \beta \end{bmatrix} where\, U = H, X$$

U is called the generalized single qubit gate given the constraint that U must be unitary.

Tip   A matrix U is unitary if multiplied by its Hermitian transpose $U^\dagger$ gives the identity matrix: $U^\dagger U = I$. The Hermitian transpose or conjugate transpose is denoted by a dagger (†) symbol $U^\dagger = (U^T)^*$ that is the complex conjugate of the transposed.

The transpose of a matrix is a new matrix whose rows are the columns of the original. For example, if $A = \begin{bmatrix} a & b \\ c & d \end{bmatrix}$ then $A^T = \begin{bmatrix} a & c \\ b & d \end{bmatrix}$. Then, to obtain the Hermitian transpose $A^\dagger = \begin{bmatrix} a & c \\ b & d \end{bmatrix}^*$, take the complex conjugate of each entry. (The complex conjugate of a + bi, where a and b are reals, is a – bi, that is switch the sign of the imaginary part if any).

Note that both gates H and X must be unitary. This can be easily verified by calculating $X^\dagger X = I$ and $H^\dagger H = I$:

$$X = \begin{bmatrix} 0 & 1 \\ 1 & 0 \end{bmatrix} \quad X^\dagger = \begin{bmatrix} 0 & 1 \\ 1 & 0 \end{bmatrix} \rightarrow X^\dagger X = XX = I$$

$$H = \frac{1}{\sqrt{2}} \begin{bmatrix} 1 & 1 \\ 1 & -1 \end{bmatrix} \quad H^\dagger = \frac{1}{\sqrt{2}} \begin{bmatrix} 1 & 1 \\ 1 & -1 \end{bmatrix} \rightarrow H^\dagger H = HH = I$$

Unitary Matrices Are Good for Quantum Gates

A question that arises from the previous section: Why go through all the trouble above? Why do X and H need to be unitary? The answer is that unitary matrices preserve vector length. This is useful for quantum gates because quantum gates require input and output states to be normalized (have a vector length of one). In fact unitary matrices are the only type of matrices that preserve length and therefore the only type of matrix that can be used for quantum gates. All in all, a deeper question arises, why do quantum gates should be linear in the first place and why use a matrix representation at all? We'll try to answer this in a later section, but for now, we'll just have to accept it.

Other Single Qubit Gates

In the previous section we saw the single qubit gates X and H. At the same time, there are other single qubit gates that are useful in quantum computation.

The X gate has two partners Y, Z. These form the trio known as the Pauli Sigma (σ) gates.

q[0] |0) — X — Y — Z —

$$X = \begin{bmatrix} 0 & 1 \\ 0 & 1 \end{bmatrix}, Y = \begin{bmatrix} 0 & -i \\ i & 1 \end{bmatrix}, Z = \begin{bmatrix} 1 & 0 \\ 0 & -1 \end{bmatrix}$$

These three matrices are useful for information processing tasks such as super dense coding, a process that seeks to store classical information efficiently in a qubit. They also come up when analyzing atomic properties such as electron spin. Plus they are closely related to the three dimensions of space XYZ.

The rotation gate

q[0] |0) — T —

$$\begin{bmatrix} \cos\theta & -\sin\theta \\ \sin\theta & \cos\theta \end{bmatrix}$$

It is the familiar rotation on real space by an angle θ. This is a unitary matrix, and in this particular case the T gate performs a $\Pi/4$ rotation around the Z-axis. This gate is required for universal control.

Gates can also manipulate many qubits as we'll see in the next section.

Qubit Entanglement with the Controlled-NOT Gate

This gate completes the arsenal of quantum gates required for quantum computation. The controlled NOT (CNOT) is a two qubit gate with four computational basis states.

For a superposition the four basis states CNOT gives:

$$\alpha|00\rangle + \beta|01\rangle + \delta|10\rangle + \gamma|11\rangle$$

where (alpha) α, (beta) β, (delta) δ, and (gamma) γ are the superposition coefficients. The quantum circuit is shown as follows:

q[0] |0⟩ ———⊕———

q[1] |0⟩ ———●———

The matrix representation of CNOT for the basis states is given by:

$$
\begin{bmatrix}
1 & 0 & 0 & 0 \\
0 & 1 & 0 & 0 \\
0 & 0 & 0 & 1 \\
0 & 0 & 1 & 0
\end{bmatrix}
\begin{matrix}
|00\rangle \\
|01\rangle \\
|10\rangle \\
|11\rangle
\end{matrix}
$$

The plus (+) symbol is called the target qubit, and the blue dot (below it) is the control qubit. What it does is simple:

- If the control qubit is set to 1, then it flips the target qubit.
- Otherwise, it does nothing.

To be more precise if the first bit is the control then:

$|00\rangle \rightarrow |00\rangle$ *control* 0 *do nothing*

$|01\rangle \rightarrow |01\rangle$ *control* 0 *do nothing*

$|10\rangle \rightarrow |11\rangle$ *control* 1 *flip 2nd*

$|11\rangle \rightarrow |10\rangle$ *control* 1 *flip 2nd*

An easy representation of the above is

$|xy\rangle \rightarrow |\, x\, y \oplus x\rangle$

Tip   The CNOT gate is required to generate entanglement, and it is critical in all kinds of tasks including quantum teleportation, super dense coding, and almost any quantum algorithm out there.

For example, to entangle 2 qubits, apply the Hadamard gate (H) to the first qubit and then apply the CNOT to the second qubit as shown in the following:

q[1] |0⟩ ———⊕———

q[2] |0⟩ —[H]—●———

For the basis state in qubit (2) the Hadamard gives:

$$|00\rangle \rightarrow \frac{|00\rangle + |10\rangle}{\sqrt{2}}$$

After applying the cNOT, we flip the second qubit if the control is 1, thus:

$$|00\rangle \rightarrow \frac{|00\rangle + |11\rangle}{\sqrt{2}}$$

This effectively creates an entangled state between qubits 1 and 2.

All in all, CNOT and single qubit gates are a powerful arsenal for quantum computation. Because they build up unitary operations on any number of qubits, they are said to be universal for quantum computation. This means that to build a quantum computer that can solve any quantum task, it is enough to use single qubit gates along with CNOT and measurement gates.

Universal Quantum Computation Delivers Shortcuts over Classical Computation

You may wonder how all the circuits and algebra above can help in solving computation tasks that can be easily performed, and probably cheaper, in a classical system. If you consider the so-called bit strength of a classical system

$$x \rightarrow f(x)$$

When given some input x, the goal is to compute a function f(x) with at least 2^{k-1} elementary operations (where k is the bit strength). Then universal quantum computation can provide an equivalent circuit of roughly the same size that contains the same classical model:

$$|x,0\rangle \rightarrow |x, f(x)\rangle$$

What is exciting about the circuit above is that there are sometimes shortcuts provided by quantum computation that get results faster. This means that you can compute f(x) in fewer than 2^{k-1} operations. For some quantum algorithms such as factorization, the speedups are exponential. This is a brand new algorithmic paradigm with very few implementations out there where the possibilities are endless. Let's finish up with a review of these concepts and a set of practice exercises.

Gate Identity Cheat Sheet

These series of identities are useful for circuit optimization. Remember that quantum gates have a noise attached to them, which accrues as the circuit complexity increases; thus, reducing their number is important.

HXH = Z

Sandwich the X gate between two Hadamards to obtain Z. Thus

$$Z = \frac{1}{\sqrt{2}}\begin{bmatrix} 1 & 1 \\ 1 & -1 \end{bmatrix}\begin{bmatrix} 0 & 1 \\ 1 & 0 \end{bmatrix}\frac{1}{\sqrt{2}}\begin{bmatrix} 1 & 1 \\ 1 & -1 \end{bmatrix} = \begin{bmatrix} 1 & 0 \\ 0 & -1 \end{bmatrix}$$

HZH = X

The same rule applies to the Z gate:

$$X = \frac{1}{\sqrt{2}}\begin{bmatrix} 1 & 1 \\ 1 & -1 \end{bmatrix}\begin{bmatrix} 1 & 0 \\ 0 & -1 \end{bmatrix}\frac{1}{\sqrt{2}}\begin{bmatrix} 1 & 1 \\ 1 & -1 \end{bmatrix} = \begin{bmatrix} 0 & 1 \\ 1 & 0 \end{bmatrix}$$

Control-Z

q_0 ──────●──────

q_1 ─[H]─⊕─[H]─

$$CZ = (I \otimes H)(CX)(I \otimes H) = \begin{bmatrix} 1 & 0 & 0 & 0 \\ 0 & 1 & 0 & 0 \\ 0 & 0 & 1 & 0 \\ 0 & 0 & 0 & -1 \end{bmatrix}$$

Generally, we can transform a single CNOT into a controlled version of any rotation around the Bloch sphere by an angle Pi by simply preceding and following it with the correct rotations.

Control-Y

q_0 ──────●──────

q_1 ─[$S^\dagger$]─⊕─[S]─

Control-H

q_0 ──────●──────

q_1 ─[R_Y $_{\pi/4}$]─⊕─[R_Y $_{-\pi/4}$]─

The S gate rotates by $\pi/2$ over the Z-axis of the Bloch sphere:

$$S = \begin{bmatrix} 1 & 0 \\ 0 & i \end{bmatrix}, S^\dagger = \begin{bmatrix} 1 & 0 \\ 0 & -i \end{bmatrix}$$

$$RY(x) = \begin{bmatrix} \cos\dfrac{x}{2} & -\sin\dfrac{x}{2} \\ \sin\dfrac{x}{2} & \cos\dfrac{x}{2} \end{bmatrix}, x = \frac{\pi}{4}$$

This is a doozy: The SWAP-Gate can be represented by 3 CNOTs:

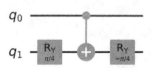

$$SWAP = (ZC)(CZ)(ZC) = \begin{bmatrix} 1 & 0 & 0 & 0 \\ 0 & 0 & 1 & 0 \\ 0 & 1 & 0 & 0 \\ 0 & 0 & 0 & 1 \end{bmatrix}$$

Exercise 5.20: Show by matrix multiplication that SWAP = (ZC)(CZ)(ZC). Hint: The Z gate is its own inverse; thus, ZC = CZ.

$$SWAP = (ZC)(CZ)(ZC) = \begin{bmatrix} 1 & 0 & 0 & 0 \\ 0 & 1 & 0 & 0 \\ 0 & 0 & 1 & 0 \\ 0 & 0 & 0 & -1 \end{bmatrix} \begin{bmatrix} 1 & 0 & 0 & 0 \\ 0 & 1 & 0 & 0 \\ 0 & 0 & 1 & 0 \\ 0 & 0 & 0 & -1 \end{bmatrix} \begin{bmatrix} 1 & 0 & 0 & 0 \\ 0 & 1 & 0 & 0 \\ 0 & 0 & 1 & 0 \\ 0 & 0 & 0 & -1 \end{bmatrix} = \begin{bmatrix} 1 & 0 & 0 & 0 \\ 0 & 0 & 1 & 0 \\ 0 & 1 & 0 & 0 \\ 0 & 0 & 0 & 1 \end{bmatrix}$$

Exercise 5.21: Which of the following quantum gates resembles the probability of classical coin flip: X, Z, Z, H?

Quantum Gate vs Boolean Gate Cheat Sheet

Boolean gates have their quantum counterparts which can be constructed using the so-called Clifford set (X, Y, Z, H, Toffoli, and S). Here is another very useful set of identities you should remember.

Boolean AND is equivalent to Toffoli (CCX).

$$CCX = \begin{bmatrix} 1 & 0 & 0 & 0 \\ 0 & 1 & 0 & 0 \\ 0 & 0 & 0 & 1 \\ 0 & 0 & 1 & 0 \end{bmatrix}$$

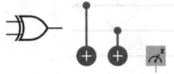

Boolean XOR can be constructed with 3 qubits and 2 CX.

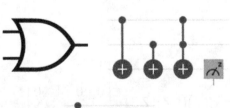

Boolean OR is made of an XOR plus a Toffoli (AND) gate.

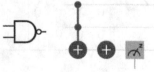

The NAND gate is very important because any boolean function can be implemented by using a combination of NAND gates. This property is called functional completeness. NAND is made of 3 qubits with Toffoli and CX in the last qubit.

Exercise 5.22: The Half Adder is a useful electronic circuit to add two single binary digits and provide the output plus a carry value. It has two inputs, A and B, and two outputs S (sum) and C (carry). The circuit and truth table is shown as follows.

A	B	C	S
0	0	0	0
0	1	0	1
1	0	0	1
1	1	1	0

Write a quantum circuit to implement the Half Adder. Use the identities in the previous section. Hint: Your circuit will have 4 qubits (A, B, S, C) with S = A⊕B and C = A AND B.

Exercises

Sharpen your knowledge of complex numbers and linear algebra even further with this extended set of exercises.

5.23 What is the complex conjugate z* of z = a+ib?

5.24 Calculate the sum of a + ib and c + id.

5.25 Calculate the product of a + ib and c + id.

5.26 What is the magnitude |z| of a complex z?

5.27 Calculate the product of $\begin{bmatrix} a & b & c \end{bmatrix} \begin{bmatrix} x0 & x1 & x2 \\ y0 & y1 & y2 \\ z0 & z1 & z2 \end{bmatrix}$

5.28 Which of the following gates act as Hermitian operators? (select 3): X, S, H, T, Z. Tip: Gates that are their own unitary inverses are called Hermitian operators.

5.29 Gates X, Y, and Z perform rotations on a Bloch sphere around the X-, Y-, and Z-axis, respectively. By which angle are these rotations performed?

It is important to have a solid mathematical background if one is to program a quantum computer. At the end, the magic of quantum boils down to clever linear algebra. Like the man hiding behind the curtain of the almighty wizard. Don't be frightened when you hear the phrase quantum mechanics. It may sound and look spooky, but at the core is just linear algebra. Now, let's set aside the maths and start coding our first quantum program.

## CHAPTER 6

Qiskit, Awesome SDK for Quantum Programming in Python

In this chapter, you will get started with Qiskit, the top SDK out there for quantum programming. You will learn how easy it is to install it in your local system. This section also shows how quantum computation can mirror its classical counterpart and find shortcuts to get results even faster. Next, the chapter walks through the anatomy of a quantum program including system calls, circuit compilation and design, quantum assembly, and more.

Qiskit packs a set of helpful simulators to execute your programs locally or remotely, but it also allows you to run in a real processor. Step by step, you will learn how to run your quantum programs in a real device within the IBM Quantum cloud platform. So let's get to it.

Installing Qiskit

Qiskit is the Quantum Information Software Kit, the de facto SDK for quantum programming in the cloud. It is written in Python, a powerful scripting language for scientific computing. Let's see how the SDK can be installed in Linux or Windows systems. We'll begin with the easiest (Windows) and then jump to Linux (CentOS, Ubuntu).

© Vladimir Silva 2024
V. Silva, *Quantum Computing by Practice*, https://doi.org/10.1007/978-1-4842-9991-3_6

Setting Up in Windows

Qiskit requires Python 3.6 or later. If you have a Windows system, chances are that you don't have Python installed. If so, you can get the installers from the Python.org website. Download the installer, run it, and verify your installation by running the following from the command window:

```
C:\>Python -V
Python 3.8.10
```

Wrangling Python versions can be confusing. Nevertheless, it is recommended that you use a version above 3.8 to access the latest features of the SDK. Python features an amazing package manager called PIP (preferred installer program) which makes installing modules very easy. Thus, to install Qiskit simply type at the console:

```
C:\>pip install qiskit qiskit-aer qiskit-ibmq-provider
```

The packages you need to get started include qiskit (core libraries), qiskit-aer (simulators and device libraries), and qiskit-ibmq-provider (for access to the hardware processors). You can see the list of packages installed in your host by typing: *pip list*. Here is a sample of my configuration (Listing 6-1).

Listing 6-1. Python package information for Qiskit SDK

```
qiskit                  0.39.0
qiskit-aer              0.11.0
qiskit-experiments      0.3.1
qiskit-ibmq-provider    0.19.2
qiskit-metal            0.1.2
qiskit-nature           0.6.0
```

That is it; you have taken the first step in this journey as a quantum programmer. For the Linux user, let's set things up in CentOS 7.

Setting Up in Linux CentOS

Things are a bit trickier to setup in CentOS 7. This is because CentOS focuses mainly on stability than bleeding edge software. Thus, CentOS comes with Python 2.7 out of the box; furthermore, the official distribution does not provide packages for Python 3.6. This doesn't mean however that Python 3.6 cannot be installed. Let's see how.

Tip   The instructions in this section should work for any Linux flavor based on the Red Hat base such as RHEL 7, CentOS7, and Fedora Core.

Step 1: Prepare Your System

First, make sure that Yum (The Linux Update Manager) is up to date by running the command:

```
$ sudo yum -y update
```

Next, `install yum-utils`, a collection of utilities and plugins that extend and supplement yum:

```
$ sudo yum -y install yum-utils
```

Install the CentOS Development Tools. These include compilers and libraries to allow for building and compiling many types of software:

```
$ sudo yum -y groupinstall development
```

Now, let's install Python 3. Note that we'll run multiple versions of Python: The official, 2.7 and 3.6 for development.

Step 2: Install Python 3

To break out of the chains of the default CentOS distribution, we can use a community project called *Inline with Upstream Stable* (IUS). This is a set of the latest development libraries for OSes that don't provide them such as CentOS. Let's install IUS through yum:

```
$ sudo yum -y install https://centos7.iuscommunity.org/ius-release.rpm
(CentOS7)
```

Once IUS is finished installing, we can install the most recent version of Python (3.6):

```
$ sudo yum -y install python36u
```

Check to make sure that the installation is correct:

```
$ python3.6 -V
Python 3.6.4
```

Now, let's install pip and verify:

```
$ sudo yum -y install python36u-pip
$ pip3.6 -V
```

Finally, we will need to install the IUS package python36u-devel, which provides useful Python development libraries:

```
$ sudo yum -y install python36u-devel
```

Step 3: Don't Disturb Others – Set Up a Virtual Environment

This step is useful only if you run Linux and have a multiuser system running multiple versions of Python and don't want to disturb other users. For example, to create a virtual environment in your home folder:

```
$ mkdir $HOME/qiskit
$ cd $HOME/qiskit
$ python3 -m venv qiskit
```

The command sequence above creates a folder called qiskit in the user's home to contain all your quantum programs. Inside this folder, a virtual Python 3 environment called qiskit is also created. To activate the environment run the command:

```
$ source qiskit/bin/activate
(qiskit) [centos@localhost qiskit]$
```

Within the virtual environment, you can use the command python instead of python3, and pip instead of pip3 if you prefer:

```
$ python3 -V
Python 3.9.4
```

Tip   If you don't activate your virtual environment, then you must use python3 and pip3 instead of python and pip. Also keep in mind that things may be different depending on your Linux flavor.

Step 4: Install Qiskit

Activate your virtual environment and install Qiskit with the command:

```
$ pip install qiskit qiskit-aer qiskit-ibmq-provider matplotlib
```

Listing 6-2 shows the standard output of the preceding command.

Listing 6-2. Qiskit installation in CentOS 7

```
Collecting qiskit
  Downloading qiskit-0.5.7.tar.gz (4.5MB)
    100% |████████████████████████████████| 4.5MB 183kB/s
Collecting matplotlib<2.2,>=2.1 (from qiskit)
  Downloading matplotlib-2.1.2.tar.gz (36.2MB)
    100% |████████████████████████████████| 36.2MB 18kB/s
    Complete output from command python setup.py egg_info:
    ======================================================================
    Edit setup.cfg to change the build options
...
Installing collected packages:  IBMQuantum, numpy, python-dateutil, pytz,
cycler, pyparsing, matplotlib, decorator, networkx, ply, scipy, mpmath,
sympy, pillow, qiskit
  Running setup.py install for pycparser ... done
  Running setup.py install for matplotlib ... done
  Running setup.py install for networkx ... done
  Running setup.py install for ply ... done
  Running setup.py install for mpmath ... done
  Running setup.py install for sympy ... done
```

```
  Running setup.py install for qiskit ... done
Successfully installed IBMQuantum-1.9.0  qiskit-0.4.11 requests-2.18.4
requests-ntlm-1.1.0 scipy-1.0.1 six-1.11.0 sympy-1.1.1 urllib3-1.22

(qiskit) [centos@localhost qiskit]$
```

Tip Under a virtual environment, Python packages will be installed in the
environment's home lib/python3.6/site-packages instead of the system's path as
shown in Figure 6-1.

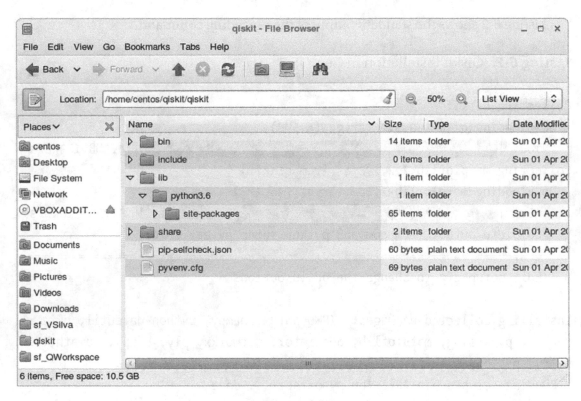

Figure 6-1. *Python virtual environment folder layout*

Credentials Configuration

This step is required to run your programs on hardware or a remote simulator. It involves storing your API token in local disk under your home directory, so the SDK can talk to the cloud. As a matter of fact, Qiskit uses the REST API described in Chapter 4. Configuration can be done programmatically, by typing the following lines in the python interpreter from the command line:

```
from qiskit import IBMQ
IBMQ.save_account('MY_API_TOKEN')
```

It also can be done manually:

1. From the IBM Quantum dashboard https://quantum-computing.ibm.com/, copy the API token.

2. In your home directory, add the token to the file $HOME/.qiskit/qiskitrc.

    ```
    [ibmq]
    token = API-TOKEN
    url = https://auth.quantum-computing.ibm.com/api
    verify = True
    ```

3. Add the token to the file $HOME/.qiskit/qiskit-ibm.json.

    ```
    {
        "default-ibm-quantum": {
            "channel": "ibm_quantum",
            "token": "API-TOKEN",
            "url": "https://auth.quantum-computing.ibm.com/api"
        }
    }
    ```

Note to Linux users   The permissions should be 755 for the folder and 644 for all files.

We are now ready to start writing quantum code. Let's see how.

Your First Quantum Program

Let's look at the anatomy of a quantum program with a bare bones example. In this example we create a single qubit, 1 classic register to measure the qubit, then we apply the Pauli X gate (bit flip) on the qubit and finally measure its value. The basic pseudo code of the program can be resumed as follows:

1. Create a quantum program.

2. Create one or more qubits and classical registers to measure the qubits.

3. Create a circuit which groups the qubits in a logical execution unit.

4. Apply quantum gates on the qubits to achieve the desired result.

5. Measure the qubits into the classical register to collect a final result.

6. Execute in the simulator or real quantum device.

7. Fetch the results.

Now let's look at the Python code as well as the composer circuit in detail.

Listing 6-3. Anatomy of a quantum program

```
##############################
from qiskit import *
from qiskit.tools.visualization import *

def main():
    # create a 1 qubit  circuit with  1 classic register
    qc = QuantumCircuit(1,1)

    # Pauli X gate
    qc.x(0)
```

```
# measure gate from qubit 0 to classical bit 0
qc.measure(0, 0)

# Print in studout
print(qc)

# backend simulator
backend = 'qasm_simulator'

# run in simulator
job = execute(qc, Aer.get_backend(backend))

# Show result counts
print (job.result().get_counts())

if __name__ == '__main__':
    main()
```

Let's see what is going on in Listing 6-3:

- Lines 2-3 import the required libraries: qiskit (quantum classes), qiskit.tools.visualization (for circuit visualization).

- Next, line 7 creates a QuantumCircuit. This is the access point to all operations. This circuit has 1 quibit and 1 classical register to measure it.

- Line 10 adds a Pauli X gate (bit flip) to the first qubit. This will flip the initial state |0> to |1>.

- Line 13 measures the qubit to the classical register. Line 16 dumps a text graph of the circuit to standard output.

- Finally, run in the simulator qasm_simulator (lines 19-25).

Windows developers, watch out. You must wrap your program in a main function and then call it with:

```
if __name__ == '__main__':
    main()
```

This is required in Windows because Qiskit executes the program using asynchronous tasks (executors), and when the task fires, the subprocess will execute the main module at start-up. Thus, you need to protect the main code to avoid creating subprocesses recursively. I found this out the hard way when my programs run properly in CentOS but failed in Windows with:

```
RuntimeError:
        An attempt has been made to start a new process before the
        current process has finished its bootstrapping phase.

        This probably means that you are not using fork to start your
        child processes and you have forgotten to use the proper idiom
        in the main module:

            if __name__ == '__main__':
                freeze_support()
                ...

        The "freeze_support()" line can be omitted if the program
        is not going to be frozen to produce an executable.
```

This can be a source of grief for the newcomer to Python. Now, run the program to see its output:

```
python p1.py
```

```
{'1': 1024}
```

The result is the JSON document {'1': 1024} where 1 is the measurement of the qubit (remember that we used an X gate to flip the bit) and 1024 is the number of iterations (shots) of that result. The probability of this result is calculated by dividing the number of the result iterations (1024) by the total number of iterations of the program (1024). In this case $P = 1024/1024 = 1$.

Tip   Quantum computers are probabilistic machines. Thus, all measurements come attached with a probability for that specific result.

Listing 6-3 can also be described with an equivalent quantum circuit quickly constructed and executed in the Quantum composer as shown in Figure 6-2.

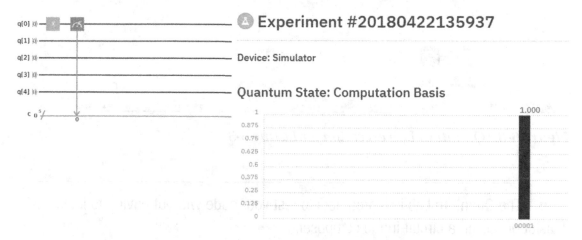

Figure 6-2.   *Composer experiment for program 6-3*

Figure 6-2 shows the quantum circuit for Listing 6-3 including the result of the experiment as well as the attached probability. The circuit is very simple as you can see: In the composer, drag an X gate over qubit 0, then perform a measurement on the same qubit. You will find the composer a wonderful tool to construct relatively simple circuits, execute them, and visualize their results.

Quantum Lab: A Hidden Jewel Within the Cloud Console

You have used the command line and the composer to run your first program, but there is a very powerful tool in the IBM console arsenal that is very useful when you need to test your program quickly. It's called the Quantum Lab.

Exercise 6.1

From the console main menu, select Quantum Lab; you will be presented with a Launcher. Select a Python 3 Notebook and paste lines 2, 3, and 6-25 from Listing 6-3 (remove all indentations). Press the Run button to see your program in action (see Figure 6-3).

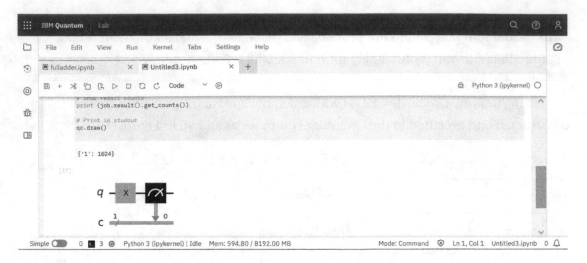

Figure 6-3. *Quantum Lab execution for Listing 6-3*

Tip   The Quantum Lab is a great tool to test your code without having to install Qiskit or design a circuit in the composer.

Exercise 6.2

Modify the program to display the circuit in the browser by invoking the Quantum Circuit draw() method to display it. (Hint: add qc.draw() at the end of the code). Verify your circuit looks like the bottom of Figure 6-3.

Exercise 6.3

Plot a histogram of the result counts. Verify it matches the output of the program and the composer. (Hint: use the plot_histogram() system call, passing the job result counts as a parameter).

Now let's peek into the SDK internals to see how this code gets massaged behind the scenes.

SDK Internals: Circuit Compilation

Figure 6-4 shows what goes on behind the scenes when your program is run:

- Qiskit compiles your program's circuit(s) into a JSON document to be submitted to the local simulator.

- The simulator parses the document, runs the circuit, and returns an opaque JSON document (hidden from the developer).

- Qiskit wraps the results JSON document in an object available to the main program: For example, a call to `result.get_counts('Circuit')` extracts the count information from this document.

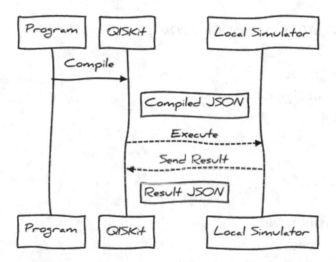

Figure 6-4. *Sequence diagram between the program, Qiskit, and local simulator*

Circuit Compilation

Listing 6-4 shows the format of the compiled program before submission to the simulator. The document is made up of

- An execution id.

- A header with information about the simulator including name, number of credits used in the execution plus number of run interactions (shots).

- The circuits section contains an array of circuit objects. Each circuit is made of

  - A circuit name.

  - A header (config) with information such as qubit coupling map, basis (physical) gates, run time seed, and more.

  - A compiled circuit section with a header containing information about the qubits and classical registers. As well as an array of operations (or gates) applied to the circuit and their parameters.

Listing 6-4. Compilation format for program 6-3

```
{
  "id": "aA46vJHgnKQko3u5L1QqbUDk31sY2m",
  "config": {
    "max_credits": 10,
    "backend": "local_qasm_simulator",
    "shots": 1024
  },
  "circuits": [{
    "name": "Circuit",
    "config": {
      "coupling_map": "None",
      "layout": "None",
      "basis_gates": "u1,u2,u3,cx,id",
      "seed": "None"
    },
    "compiled_circuit": {
      "operations": [{
        "name": "u3",
        "params": [3.141592653589793, 0.0, 3.141592653589793],
        "texparams": ["\\pi", "0", "\\pi"],
        "qubits": [0]
      }, {
```

```
      "name": "measure",
      "qubits": [0],
      "clbits": [0]
    }],
    "header": {
      "number_of_qubits": 1,
      "qubit_labels": [
        ["qr", 0]
      ],
      "number_of_clbits": 1,
      "clbit_labels": [
        ["cr", 1]
      ]
    }
  },
  "compiled_circuit_qasm": "OPENQASM 2.0;\ninclude \"qelib1.inc
  \";\nqreg qr[1];\ncreg cr[1];\nu3(3.14159265358979,0,3.14159265358979)
  qr[0];\nmeasure qr[0] -> cr[0];\n"
  }]
}
```

Note   The compilation format is opaque to the programmer and not meant to be accessed directly but via the SDK API. The reason is that its format may change from version to version. However, it is always good to understand what occurs behind the scenes.

Execution Results

This is the response document from the backend simulator to the client. The format of this document is shown in Listing 6-5. Remarkable information includes

- Status of the run, execution time, simulator name, and more.

- Result data. This is the information available within your program including circuit name, execution time, status, measurement shots, and more.

Listing 6-5. Results document from local simulator

```
{
  "backend": "qasm_simulator",
  "id": "aA46vJHgnKQko3u5L1QqbUDk31sY2m",
  "result": [{
    "data": {
      "counts": {
        "1": 1024
      },
      "time_taken": 0.0780002
    },
    "name": "Circuit",
    "seed": 123,
    "shots": 1024,
    "status": "DONE",
    "success": true,
    "threads_shot": 4
  }],
  "simulator": "qubit",
  "status": "COMPLETED",
  "success": true,
  "time_taken": 0.0780002
}
```

Obtaining the results document is a bit trickier because it is an opaque object not exposed to the user's program. The important thing to remember is that the results document (as well as the compilation format) is hidden from the programmer. The reason is that their formats may change over time; nonetheless, it is always helpful to understand how things work behind the scenes.

Tip   The compilation and results formats are useful for simulator and integration developers. For example, you could add quantum functionality to your organization web applications.

Assembly Code

The compiled circuit in Listing 6-4 includes a section that contains a translation of the program into Quantum Assembly (QASM) as shown in the next paragraph.

```
OPENQASM 2.0;
include "qelib1.inc";
qreg qr[1];
creg cr[1];
x qr[0];
measure qr[0] -> cr[0];
```

Tip   QASM is useful only if running in the remote simulator provided by IBM Quantum.

Qiskit Simulators

Access to real quantum devices comes at a premium; thus, we shouldn't run trivial programs such as Listing 6-3 in real devices. For testing purposes, Qiskit packs a small army of simulators to satisfy all your needs. Table 6-1 provides a list of some of them available at the time of this writing.

Table 6-1. *List of local and remote simulators for IBM Quantum*

Name	Description
qasm_simulator	This is the default Python simulator bundled with Qiskit. It is slow but does the job
simulator_mps	Matrix product state, an efficient classical simulation of entangled states
simulator_statevector	A Statevector Simulator supports CPU and GPU simulation methods
Complete list.	Check out the full list at https://qiskit.org/ecosystem/aer/tutorials/1_aer_provider.html

As a simple exercise, obtain a list of IBM Quantum simulators and real devices by pasting the following REST API URL into your browser:

```
https:// api-qcon.quantum-computing.ibm.com/api/Backends?access_
token=ACCESS_TOKEN.
```

Note that you need an access token which can be easily obtained using the REST API described in Chapter 4. Now that you have learned how to run a program in the simulator, let's do it in the real thing.

Running in a Real Quantum Device

There are three ways of running in a real device from easiest to hardest:

1. Use the composer to create your circuit and quickly run on hardware. This method is mostly for didactic purposes to test things quickly.

2. Use Python on your local desktop to execute your program. This method is for serious quantum programming and algorithm development.

3. Use the REST API: This method can be used if you work for a service provider and wish to integrate quantum into your web or desktop platform.

Let's modify the program from the previous section to make a more complex circuit instead. Listing 6-6 shows a circuit that performs a series of rotations on the first qubit. The rotations demonstrate the use of the physical U-gate to rotate a single qubit over the X-, Y-, and Z-axis of the Bloch sphere by theta, phi, or lambda degrees.

Tip   Physical gates (also known as basis gates) are important because they constitute the foundation under which more complex logical gates are constructed.

Listing 6-6 performs the following steps:

*   Allocates 5 qubits and 5 classical measurement registers corresponding to the 5 qubits available from the quantum processor ibm_perth form Quantum (line 8).

- Next, a sequence of rotations on the first qubit are performed using the basis gates u1, u2, and u3 (lines 29-34).

- Finally, a measurement is performed in the qubit, and the result is stored in the classical register.

- Before execution the backend is set to ibmqx4 (a 5 qubit processor – line 42), and the authentication token and API URL are set via `set_api(Qconfig.APItoken, Qconfig.config['url'])`.

- To execute in the real quantum device use the execute system call `execute(NAMES, BACKEND, shots=SHOTS, timeout=TIMEOUT)` where

  - NAMES is a list of circuit names.

  - SHOTS is the number of iterations performed in the circuit. The higher the number, the greater the accuracy.

  - TIMEOUT is the read timeout from the remote end point.

Listing 6-6. Sample circuit #2

```
import sys,time,math
from qiskit import *
from qiskit.tools.visualization import *

# Main sub
def main():
  # Circuit with 5 qubits, 5 classic bits
  circuit = QuantumCircuit (5,5)

  # first physical gate: u1(lambda) to qubit 0
  circuit.u(math.pi/2, -4 *math.pi/3, 2 * math.pi, 0)
  circuit.u(math.pi/2, -3 *math.pi/2, 2 * math.pi, 0)
  circuit.u(-math.pi, 0, -math.pi, 0)
  circuit.u(-math.pi, 0, -math.pi/2, 0)
  circuit.u(math.pi/2, math.pi, -math.pi/2, 0)
  circuit.u(-math.pi, 0, -math.pi/2, 0)

  # measure gate from qubit 0 to classical bit 0
  circuit.measure_all()
```

```
  # dump to stdout
  print (circuit)

  # HW
  name       = 'ibm_perth'

  IBMQ.load_account()
  provider   = IBMQ.get_provider(hub='ibm-q', group='open',
project='main')
  backend    = provider.get_backend(name)

  # Group of circuits to execute
  circuits = [circuit]

  job = execute(circuits, backend, shots=512)

  # Show result counts
  print ("Job id=" + str(job.job_id()) + " Status:" + str(job.status()))

###########################################
if __name__ == '__main__':
  start_time = time.time()
  main()
  print("--- %s seconds ---" % (time.time() - start_time))
```

Run via composer

The program in Listing 6-6 can also be created in the Quantum composer using its slick drag and drop user interface. Simply drag the gates into the qubit histogram as shown in Figure 6-5, set the parameters for the gate(s), and finally save and run in the simulator or real device.

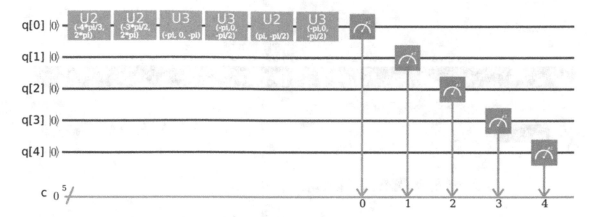

Figure 6-5. *Quantum composer Circuit for program 6-6*

For those of you who prefer the raw power of assembly, the composer allows to copy-paste code directly into the console in Assembly mode as shown in Figure 6-6. It will even parse any syntax errors in your code and show you the offending line(s).

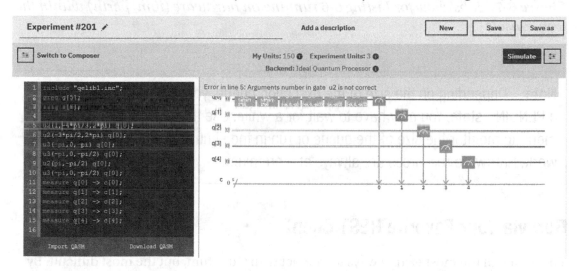

Figure 6-6. *Composer in assembly mode for circuit in Figure 6-5*

There are multiple ways of executing your experiment in IBM Quantum; one of the most interesting is using their awesome REST API.

Run via Your Local Desktop

This method is for the python programmer within you. Run the program in Listing 6-6, then monitor its status using the IBM Jobs menu from the web console (see Figure 6-7).

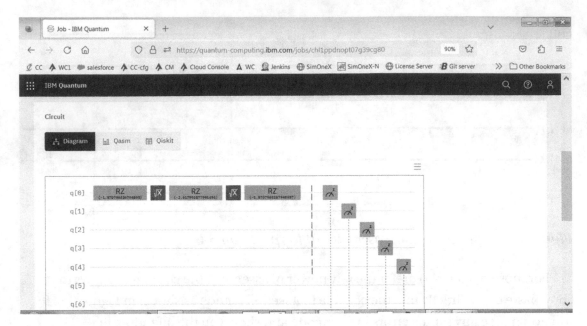

Figure 6-7. Jobs' view for Listing 6-6 running on hardware (ibm_perth) within the web console

---

Note   Depending on the size of the execution queue, your job will surely enter a PENDING state. You may have to wait for a while. Use the Compute Resources menu to monitor the size of the queue or run in the simulator until you finalize your work. Hardware resources are always at a premium.

Run via Your Favorite REST Client

This is one of the most exiting ways to interact with Quantum, but the most difficult. By using simple REST requests, you can do pretty much anything:

- List backend devices.

- List hardware or calibration parameters for the real devices.

- Get information about the job execution queue.

- Get the status of a job or experiment.

- Run or cancel jobs.

Tip   The REST API allows you to use any language to create your own interface to IBM Quantum (even a web browser). This API is described in full detail in Chapter 4. As a matter of fact, Python and the composer invoke the REST API behind the scenes.

To submit an experiment using REST, we use the jobs API. Let's see how.

Run via the Jobs API

You can use your favorite browser REST client to submit the experiment in Listing 6-6. For example, using Chrome's YARC (Yet Another REST Client), create an HTTP POST request to the endpoint:

```
https://runtime-us-east.quantum-computing.ibm.com/jobs
```

The tricky part is getting your access token or access key. For this part you must authenticate using your API token or user name and password. Note that the API token is not to be confused with the access token. To obtain an access token, you must do an authentication request. (Take a look at Chapter 4 under *Remote REST API*.)

Tip   Chrome's YARC allows you to construct REST requests and save them as favorites. Create an authentication request to IBM Quantum as described in Chapter 4, save it as a favorite, and use it every time to obtain an access token to test other REST API calls.

The request payload is a JSON document shown in Listing 6-7. The format is described in Table 6-2.

Table 6-2. *Request format for the Jobs API*

Key	Description
Circuits	This is an array of assembly code programs all in one line separated by the line feed character (\n)
shots	The number of iterations you code will go through
backend	This is an object that describes the backend. In this case ibm_quito
Hub, Group, Project	For the open (free) plan: ibm-q, open, main

Listing 6-7. HTTP Request for the Jobs API

```
{
  "program_id": "circuit-runner",
  "hub": "ibm-q",
  "group": "open",
  "project": "main",
  "backend": "ibmq_quito",
  "params": {
    "shots": 1024,
    "circuits": [
        "\n\ninclude \"qelib1.inc\";\nqreg q[5];\ncreg c[5];\
        nu2(-4*pi/3,2*pi) q[0];\nu2(-3*pi/2,2*pi) q[0];\nu3(-pi,0,-pi)
        q[0];\nu3(-pi,0,-pi/2) q[0];\nu2(pi,-pi/2) q[0];\nu3(-pi,0,-pi/2)
        q[0];\nmeasure q -> c;\n"
    ]
  },
  "tags": [
      "composer-info:composer:true",
      "composer-info:code-id:d2262dd4e07a9f894caabd977b10c7e98d90537228723
      dd14dceaab023c36098",
      "composer-info:code-version-id:646d2a856260a179aa094d3e"
  ]
}
```

Once you have obtained an access token, add the HTTP header **X-Access-Token: ACESS_TOKEN** to the request, then copy-paste the payload from Listing 6-7 into the REST client payload, submit, and wait for a response. If all goes well, you should get the following response:

```
{
    "id": "chobhqanajhpa6495on0",
    "backend": "ibmq_quito"
}
```

The Job id can be used to inspect the status or retrieve results from the console.

Tip   The Jobs API is undocumented and not meant to be accessed directly at this point. Thus, the response format may vary over time. Perhaps, this will change in the future, and the REST API will be part of the official SDK. In the meantime however your results may be different.

Result Visualization Types

Qiskit packs a sophisticated visualization library to display your results. Here are some of the types:

- Histogram: By far the most popular. It draws a histogram of the counts data.

- State City: It draws the cityscape of a quantum state; that is, two 3D bar graphs of the real and imaginary parts of the density matrix rho.

- State Hinton: It represents the values of a matrix using squares, whose size indicates the magnitude of their corresponding value and their color encodes its sign. A white square means the value is positive and a black one means negative.

- Quantum Sphere (QSphere): It plots a sphere representation of a quantum state. Here, the size of the points is proportional to the probability of the corresponding term in the state, and the color represents the phase.

- PauliVec: It draws a bar graph of the density matrix of a quantum state using as a basis all possible tensor products of Pauli operators and identities.

- Bloch Multivector: It draws a Bloch sphere for each qubit.

Note   All visualization types except Histogram require the state vector simulator.

Let's see these diagrams in action by plotting the results for the First Bell (entangled) state $\varphi_+ = \dfrac{1}{\sqrt{2}}(|00+|11)$

```
from qiskit import *
from qiskit.
visualization import *

qc = QuantumCircuit(2)
qc.h(0)
qc.cx(0,1)
qc.measure_all()

job = execute(qc, Aer.
get_backend("qasm_
simulator"))
plot_histogram (job.
result().get_counts())
```

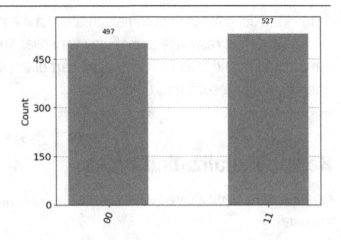

```
from qiskit import *
from qiskit.
visualization import *

qc = QuantumCircuit(2)
qc.h(0)
qc.cx(0,1)

job = execute(qc,
Aer.get_
backend("statevector_
simulator"))
plot_state_city
(job.result().get_
statevector())
```

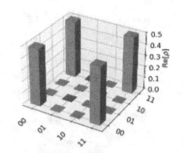

```
from qiskit import *
from qiskit.
visualization import *

qc = QuantumCircuit(2)
qc.h(0)
qc.cx(0,1)

job = execute(qc,
Aer.get_
backend("statevector_
simulator"))
plot_state_hinton(job.
result().get_
statevector())
```

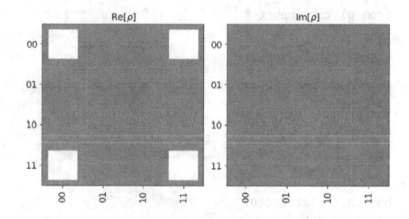

```
from qiskit import *
from qiskit.
visualization import *

qc = QuantumCircuit(2)
qc.h(0)
qc.cx(0,1)

job = execute(qc,
Aer.get_
backend("statevector_
simulator"))
plot_state_qsphere(job.
result().get_
statevector())
```

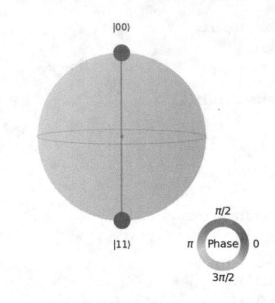

```
from qiskit import *
from qiskit.
visualization import *

qc = QuantumCircuit(2)
qc.h(0)
qc.cx(0,1)

job = execute(qc,
Aer.get_
backend("statevector_
simulator"))
plot_state_paulivec
(job.result().
get_statevector())
```

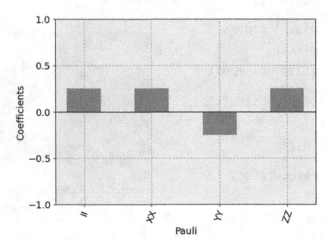

```
from qiskit import *
from qiskit.
visualization import *

qc = QuantumCircuit(2)
qc.h(0)
qc.cx(0,1)

job = execute(qc,
Aer.get_
backend("statevector_
simulator"))
plot_bloch_
multivector(job.
result().get_
statevector())
```

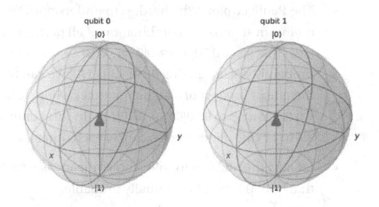

- For the histogram plot, the X-axis shows the measurement results: 00, 11 for the corresponding Bell states. The y-axis shows the result counts which amount to a probability of 1/2 (497/1024 and 527/1024 given that 497+527=1024 – the total number of shots).

- The state city plot is trickier to understand, but if you look carefully, you realize the states 00, 11 are the only ones that have a 3D bar. On the y-axis the probability is 1/2. Note that the imaginary part (right side) is empty (remember that after measurement the imaginary part will always vanish).

- The state Hinton shows white squares for the states 00, 11. The white color means the value is positive.

- The Qsphere plot shows blue circles for states 00, 11 as well; here the blue shade indicates a positive phase (sign) as shown in the legend.

- The Paulivec plot is the hardest to understand: We can rewrite the quantum state as the combination of all permutations for two qubits such that $|\psi\rangle = 0.70|00\rangle + 0|01\rangle + 0|10\rangle + 0.70|11\rangle$. Because the probability of measuring 00 or 11 is 1/2, we can represent the state as the tensor product of all Pauli matrices where $0.25I + 0.25XX - 0.25YY + 0.25ZZ$ map to $0.70|00\rangle + 0.70|11\rangle$. In this context, I maps to 00 and ZZ to 11 (the middle terms cancel each other).

- Finally, the Multivector plot shows the measurement result 00. Note that both 00 and 11 are equally probable.

Try the following practice exercises to see this more clearly. A few more result visualization exercises are available at the end of the chapter.

Exercise 6.4

Plot the histogram and state city results for the second Bell state $\varphi_- = \frac{1}{\sqrt{2}}\left(|00\rangle - |11\rangle\right)$.

Tip: Add an X gate to the first qubit. Notice that the phase (sign) does not show in the histogram but it does show in state city. Why is that? Hint: Can probabilities be negative?

Exercise 6.5

Plot the histogram and state city results for the third Bell state $\theta_+ = \frac{1}{\sqrt{2}}\left(|01\rangle + |10\rangle\right)$.

Compare your results with the diagrams above and verify the result shows the correct outputs: 01, 10. Tip: Add an X gate to the second qubit.

Noise Models and Fake Providers

So far you have seen how to display results from perfect (noiseless) simulators. However, in reality, quantum circuits are fragile and noisy. They are easily obfuscated by interactions with the environment. It is a good idea to try your circuits in a noisy simulator first, and then move to the real device for several reasons:

- Large wait times in the execution queues.

- Your work may require a large number of qubits not available on your payment plan.

- Your time constraints may require testing your work quickly, or simply you don't have the time or patience to wait hours or days on the hardware execution queue.

Whatever the reason, Qiskit provides a robust noise simulation system, and it comes in two flavors:

- Fine-grained noise control using modules from the Aer package: A high performance simulator for quantum circuits.

- Fake providers: These are backends built to mimic the behaviors of IBM Quantum hardware using system snapshots. Fake providers run in the local desktop and are named after the processor's name. For example, for processor ibm_perth, its fake counterpart is named FakePerth.

Note   We live in the age of NISQ, Noisy Intermediate Scale Quantum Computing, with error correction being one of the hottest areas of research right now. You will soon realize that noise is a big problem in experimental results, especially with large numbers of qubits where it accrues.

Let's add some noise to one of the Bell states we have studied so far. Take a look at Listing 6-8 (Figure 6-10). It creates a noise model for the Bell state $\varphi_+ = \frac{1}{\sqrt{2}}(|00\rangle + |11\rangle)$ from the previous section.

Listing 6-8. Noise model of 10% for the first Bell state φ_+

```python
from qiskit import *
from qiskit.tools.visualization import*
from qiskit.providers.aer.noise.errors import pauli_error,
depolarizing_error
from qiskit.providers.aer.noise import NoiseModel

def get_noise (p_meas, p_gate):
  error_meas = pauli_error([('X',p_meas), ('I', 1 - p_meas)])
  error_gate1 = depolarizing_error(p_gate, 1)
  error_gate2 = error_gate1.tensor(error_gate1)
```

```
    noise_model = NoiseModel()
    # measurement error is applied to measurements
    noise_model.add_all_qubit_quantum_error(error_meas, "measure")
    # single qubit gate error is applied to x gates
    noise_model.add_all_qubit_quantum_error(error_gate1, ["x"])
    # two qubit gate error is applied to cx gates
    noise_model.add_all_qubit_quantum_error(error_gate2, ["cx"])

    return noise_model

noise_model = get_noise(0.1,0.1) # 10%

qc = QuantumCircuit(2)

qc.h(0)
qc.cx(0,1)
qc.measure_all()

backend = Aer.get_backend("qasm_simulator")
job = execute(qc, backend, noise_model = noise_model)

plot_histogram(job.result().get_counts())
```

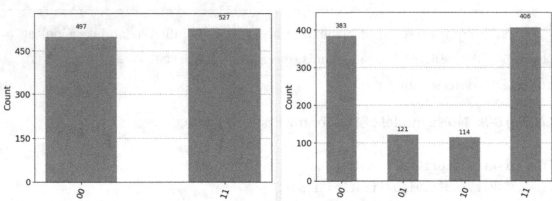

Figure 6-10. *Result counts of noiseless simulation (left) vs a noise model of 10% using the qasm_simulator from Listing 6-8 (right)*

We start by importing the required modules from the package qiskit.providers.aer. noise. Some of the key system calls include

- NoiseModel: A class that stores a noise model used for the simulation.

- pauli_error: An n-qubit Pauli error channel (mixed unitary) given as a list of Pauli Gates and probabilities.

- depolarizing_error: An n-qubit depolarizing error channel parameterized by a depolarization probability p.

Running in a fake provider is even easier; let's update the code in Listing 6-8 to use the qiskit.providers.fake_provider package for FakePerth with an error model of around 1%. Compare your result with the noiseless result from Figure 6-10 (left).

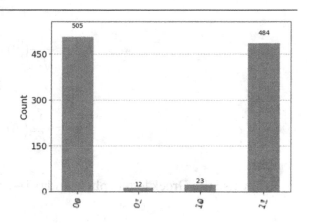

```python
from qiskit import *
from qiskit.tools.visualization import*
from qiskit.providers.fake_provider import *

qc = QuantumCircuit(2)
qc.h(0)
qc.cx(0,1)
qc.measure_all()

backend = FakePerth()

job = execute(qc, backend)
plot_histogram(job.result().get_counts())
```

Exercise 6.6

Use the Quantum Lab to create a circuit for the first GHZ state $W_+ = \frac{1}{\sqrt{2}}\left(|000\rangle + |111\rangle\right)$. Plot the histogram for noisy FakePerth. Tip: GHZ is the name for a quantum state for 3 entangled qubits. Verify the result counts are highest for 000 and 111.

Exercises

Let's create quantum circuits for the Bell states describing two entangled qubits. There are 4 Bell states:

$$\varphi_+ = \frac{1}{\sqrt{2}}\left(|00\rangle + |11\rangle\right)$$

$$\varphi_- = \frac{1}{\sqrt{2}}\left(|00\rangle - |11\rangle\right)$$

$$\varphi_+ = \frac{1}{\sqrt{2}}\left(|01\rangle + |10\rangle\right)$$

$$\varphi_- = \frac{1}{\sqrt{2}}\left(|01\rangle - |10\rangle\right)$$

They use the Greek letters phi+- and psi+-, respectively.

Exercise 6.7

Use the Quantum Lab to write a program to draw φ_+. Verify the circuit looks as shown here.

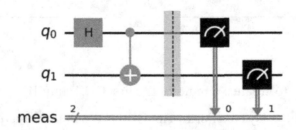

Exercise 6.8

Run the previous circuit in the qasm_simulator and display the result counts on standard output. Verify they match the expected values. For example, {'00': 517, '11': 507}. Plot a histogram of the counts. (Hint: use the plot_histogram system call).

Exercise 6.9

Modify the previous circuit to draw φ. Note that the phase (negative sign) will not show when running in the qasm_simulator. How can we visualize the negative sign? Hint: switch the qasm_simulator to the state vector simulator (statevector_simulator), then use the job.result().get_statevector() system call to print the result. Verify the negative sign shows up. Tip: the system call array_to_latex(job.result().get_statevector()) will display the output in the very nice Latex format: $[1/\sqrt{2}, 0, 0, -1/\sqrt{2}]$.

Exercise 6.10

Write a program to display the Bell state ψ+. Use the qasm_simulator. Verify the output is similar to: {'01': 517, '10': 507}. Plot the histogram.

Exercise 6.11

Finally, write a program to display the Bell state ψ-. Use the statevector_simulator. Verify the state and phase matches the algebra: $[0, 1/\sqrt{2}, -1/\sqrt{2}, 0]$.

Extended Qiskit Exercises

Here is an enhanced set to sharpen your understanding of the Qiskit SDK. Use the Quantum Lab to quickly test your solutions. If you get stuck, the answers are provided in the appendix.

6.12 Which quantum gate is similar to classical NOT gate? X, H, Y, CNOT.

6.13 Which of the following commands will convert the below quantum circuit to a qasm string?

```
qc = QuantumCircuit(2,2)
qc.h(0)
qc.h(1)
```

qc.qasm_simulator(), qc.qasm_str() , qc.qasm()

6.14 What is the depth of the following quantum circuit? Tip: The depth of a circuit is the longest path in the circuit. The path length is always an integer number, representing the number of gates it has to execute in that path.

6.15 Which of the following statements prints the qiskit version?

```
import qiskit
print(qiskit.__version__)
print(qiskit.__qiskit_version__)
print(qiskit.version())
print(qiskit_version())
```

6.16 What will be the output for the below snippet?

```
q = QuantumRegister(2,"qreg")
c = ClassicalRegister(2,"creg")
qc = QuantumCircuit(q,c)
qc.x(q[0])
qc2.measure(q,c)
job = execute(qc2,Aer.get_backend('qasm_simulator'),shots=1024)
counts = job.result().get_counts(qc2)
print(counts)
{'00': 1024}
{'11': 1024}
{'10': 1024}
{'01': 1024}
```

6.17 Which of the following statements returns the depth of the following circuit?

```
q = QuantumRegister(3)
c = ClassicalRegister(3)
qc = QuantumCircuit(q,c)
qc.h(q[0:3])
qc.x(q[0:3])
qc.z(q[0:3])
qc.draw(output='mpl')
```

```
qc.size(), qc.path(), qc.depth()
```

6.18 Which of the following options will be best suited for the missing statement in the following snippet to achieve the quantum state i|10>? Tip: Y|0> = i|1>, also remember that Qiskit uses little endian bit ordering (qubit-zero starts on the right).

```
from qiskit import QuantumRegister, ClassicalRegister, QuantumCircuit,
execute, Aer
qc= QuantumCircuit(2,2)
#missing statement
a) qc.z(1) qc.x(1)
b) qc.y(1)
c) qc.y(0)
d) qc.s(0) qc.z(1)
```

6.19 In our quantum circuit, we have a single qubit initialized to the |0> state. Which of the following quantum gates gives the same output state |0>? (select 3): S, T, HSH, HYH, I, HZH. Tip: What gate(s) leaves a state unchanged?

6.20 In the quantum circuit below, which instruction should you use to measure the qubit output? Tip: The circuit has no classical register.

```
qc = QuantumCircuit(1)
qc.x(0)
```

qc.measure(0), qc.measure(0,0), qc.measure(), qc.measure_all()

6.21 In the code below, which of the following statement is non-unitary? Tip: Ignore QuantumCircuit.

```
qc= QuantumCircuit(2,2)
qc.x(0)
qc.y(1)
qc.z(1)
qc.measure([0,1],[0,1])
```

6.22 Choose the best option to display the following plot.

```
a) legend = ['All H gates']
title= 'Superposition states of three qubits'
plot_histogram(counts, legend=legend, title=tile)
```

```
b) title = ['All H gates']
legend = 'Superposition states of three qubits'
plot_histogram(counts, legend=legend, title=tile)
```

```
c) plot_histogram(counts)
```

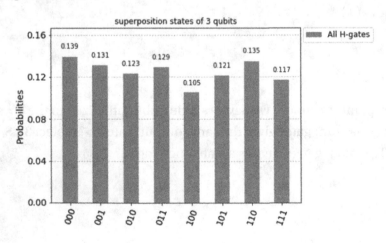

6.23 Which of the following qsphere plots given below is the correct one for the given Bell quantum circuit.

```
bell = QuantumCircuit(2)
bell.h(0)
bell.x(1)
bell.cx(0,1)
```

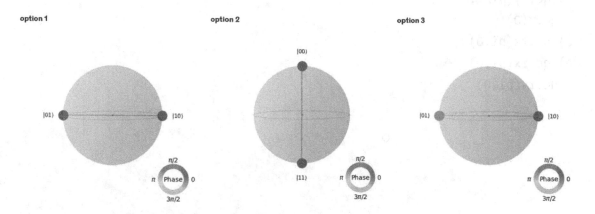

6.24 Which option describes the following given quantum circuit correctly in its state_city plot?

```
bell - QuantumCircuit(2)
bell.x(0)
bell.h(0)
bell.cx(0,1)
```

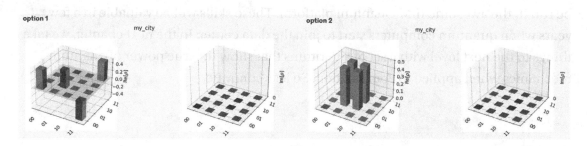

6.25 Given the state vector represented by this Bloch sphere of a single qubit quantum circuit (qc), choose the operations that lead to this state by assuming the circuit is initialized to |0> (select 3).

a) `qc.h(0)`

b) `qc.h(0)`
   `qc.x(0)`

c) `qc.ry(pi/2,0)`
   `qc.x(0)`

d) `qc.rx(pi,0)`

e) `qc.rx(pi,0)`
   `qc.ry(pi,0)`

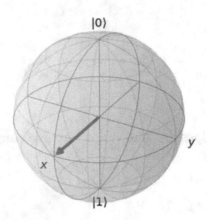

You have taken the first step in this journey as a quantum programmer in the cloud. By using the Python SDK and powerful Quantum Assembly engine, experiments can be run in the awesome IBM Quantum platform. These skills will be valuable in a few years when quantum computers start to join the data center. In the next chapter, we take things to the next level with a set of algorithms that show the true power of quantum mechanics when applied to computation. So let's continue.

Start Your Engines: From Quantum Random Numbers to Teleportation and Super Dense Coding

This chapter takes you on a journey about three remarkable information processing capabilities of quantum systems. We start with one of the simplest procedures by exploring the fundamentally random nature of quantum mechanics as a source of true randomness. Next, the chapter looks at perhaps two exuberant but related procedures called super dense coding and quantum teleportation. In super dense coding, you will learn how it is possible to send two classical bits of information using a single qubit. In quantum teleportation you will learn how the quantum state of a qubit can be recreated by a hybrid classical-quantum information transfer procedure. All algorithms include circuit design for the IBM Quantum composer as well as Python and QASM code. Results will be gathered for display and analysis, so let's get started.

Quantum Random Number Generation

In this section, you will learn how the probabilistic nature of a quantum computer can be exploited to generate random bits or numbers using the Hadamard gate.

© Vladimir Silva 2024
V. Silva, *Quantum Computing by Practice*, https://doi.org/10.1007/978-1-4842-9991-3_7

Random Bit Generation Using the Hadamard Gate

Hadamard is one of fundamental gates in any quantum information system. It is used to put a qubit in a superposition of states. Algebraically, it is described by the matrix:

$$H = \frac{1}{\sqrt{2}} \begin{bmatrix} 1 & 1 \\ 1 & -1 \end{bmatrix}$$

To understand better how this matrix puts a qubit in super position, consider the geometrical representation of a single qubit:

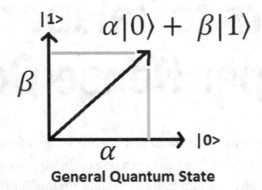

General Quantum State

Figure 7-1. *Geometric representation of the general (superimposed) state* ψ
of a qubit

In Figure 7-1, the basis states of the qubit are described using ket notation where
$|0\rangle = \begin{bmatrix} 1 \\ 0 \end{bmatrix}$ and $|1\rangle = \begin{bmatrix} 0 \\ 1 \end{bmatrix}$. Remember from the previous chapter that a ket is simply a unitary
vector (a vector of length 1). Thus, the general (or superposition) state is then defined by
the unitary vector $\psi = \alpha|0\rangle + \beta|1\rangle$ where α ab β are complex coefficients. Applying H to the
basis states gives:

$$H|0\rangle = \frac{1}{\sqrt{2}} \begin{bmatrix} 1 & 1 \\ 1 & -1 \end{bmatrix} \begin{bmatrix} 1 \\ 0 \end{bmatrix} = \frac{1}{\sqrt{2}} \begin{bmatrix} 1 \\ 1 \end{bmatrix} = \frac{1}{\sqrt{2}} \left(\begin{bmatrix} 1 \\ 0 \end{bmatrix} + \begin{bmatrix} 0 \\ 1 \end{bmatrix} \right) = \frac{|0\rangle + |1\rangle}{\sqrt{2}}$$

$$H|1\rangle = \frac{1}{\sqrt{2}} \begin{bmatrix} 1 & 1 \\ 1 & -1 \end{bmatrix} \begin{bmatrix} 0 \\ 1 \end{bmatrix} = \frac{1}{\sqrt{2}} \begin{bmatrix} 1 \\ -1 \end{bmatrix} = \frac{1}{\sqrt{2}} \left(\begin{bmatrix} 1 \\ 0 \end{bmatrix} - \begin{bmatrix} 0 \\ 1 \end{bmatrix} \right) = \frac{|0\rangle - |1\rangle}{\sqrt{2}}$$

And for the superimposed state ψ:

$$\Psi = \alpha|0\rangle + |\beta\rangle 1 \rightarrow \alpha\left(\frac{|0\rangle+|1\rangle}{\sqrt{2}}\right) + \beta\left(\frac{|0\rangle-|1\rangle}{\sqrt{2}}\right) = \frac{\alpha+\beta}{\sqrt{2}}|0\rangle + \frac{\alpha-\beta}{\sqrt{2}}|1\rangle$$

All in all, the Hadamard gate expands the range of states that are possible for a
quantum circuit. This is important because the expansion of states creates the possibility
of finding shortcuts resulting in faster computation.

Tip   Quantum mechanics says that we can't predict with certainty the values of
coefficients α, β above, even given complete knowledge of the laws of physics or a
particle's initial conditions. The best we can do is to calculate a probability.

With this in mind, a random bit generator circuit implementation is simple: In the
IBM Quantum composer, create a circuit with a Hadarmard gate for the first qubit and
then perform a measurement in the basis state as shown in Figure 7-2.

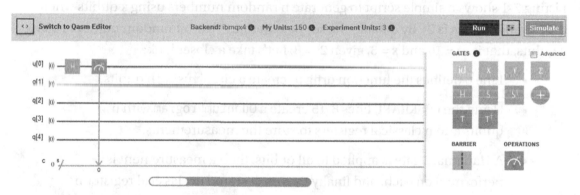

Figure 7-2. *Circuit for a random bit generation*

It is probably not a good idea to run this in the real device as it may take a while
(remember that executions are scheduled and may take time depending on the number
of jobs in the run queue). Plus each execution in a real device depletes your credits.
Run the circuit in the simulator to obtain an immediate result (see Figure 7-3). Note
that each outcome (0 or 1) has a probability of ½; thus, you can create random bits if the
probability for outcome 1 is > ½ you get a 1 else you get a 0.

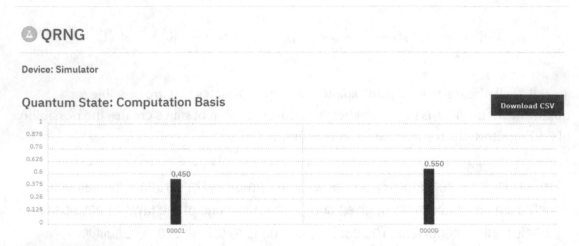

Figure 7-3. *Execution results for circuit 7-2*

Of course, this is a very inefficient way of generating random bits. A better way would
be to write a Qiskit Python script to programmatically create a circuit to do the job.
Listing 7-1 shows a simple script to generate n random numbers using x qubits where
the number of bits is 2^x. By default, the script generates 10 8-bit random numbers using
3 qubits, that is, n = 10 and x = 3, given 2^3 = 8. Let's take a closer look:

- Line 6 defines the function qrng to create a circuit using n qubits.

- Using the QISKitAPI, lines 8-15 create a `QuantumProgram` with n
  qubits and n classical registers to store the measurements.

- A Hadamard gate is applied to all qubits, then a measurement is
  performed on each, and finally the result stored in classical register n
  (lines 17-23).

- The circuit is compiled to run in the qasm_simulator. The circuit gets
  executed, and the result counts are collected (lines 26-36).

- Finally, to generate random bits look at the outcome counts. For
  example, given the results {'100': 133, '101': 134, '011': 131, '110': 125,
  '001': 109, '111': 128, '010': 138, '000': 126}. For each outcome, if the
  count is greater than the average probability, then write a 1; else write
  a zero. The average probability is calculated by dividing the number

of shots (1024 in this case) by the number of outcomes (2^x where x
is the number of qubits (default is 3) – 1024/8 = 128). Thus, for the
results above the final random binary string becomes 11100010 (226):

Listing 7-1. Quantum program to generate 100 8-bit random numbers

```
##############################
import sys,time
from qiskit import *

# Generate an 2**n bit random number where n = # of qubits
def qrng(n):

  # create n qubit(s)
  quantum_r = QuantumRegister(n, "qr")

  # create n classical registers
  classical_r = ClassicalRegister(n, "cr")

  # create a circuit
  circuit = QuantumCircuit(quantum_r, classical_r, name = "QRNG")

  # Hadamard gate to all qubits
  for i in range(n):
    circuit.h(quantum_r[i])

  # measure qubit n and store in classical n
  for i in range(n):
    circuit.measure(quantum_r[i], classical_r[i])

  # backend simulator
  backend = Aer.get_backend('qasm_simulator')

  # Group of circuits to execute
  circuits = [circuit]
  shots = 1024
  result = execute(circuits, backend, shots=shots).result()

  # Show result counts
  # counts={'100': 133, '101': 134,… , '010': 138, '000': 126}
```

```
    counts = result.get_counts()
    bits = ""

    # convert the random count to binary
    for v in counts.values():
      if v > shots/(2**n) :
        bits += "1"
      else:
        bits += "0"
    return int(bits, 2)


############################################
if __name__ == '__main__':
    start_time = time.time()
    numbers = []

    # generate 10 8 bit random numbers
    size = 100
    qubits = 3 # bits = 2**qubits
    for i in range(size):
      n = qrng(qubits)
      numbers.append(n)
    print (str(numbers) .replace('[','').replace(']',''))
```

Note   Before executing a program, always make sure your account is configured. Your account API token must exist in the file $HOME/.qiskit/qiskitrc. This is a major source of headaches. If you use the Qiskit runtime, you also need to update the file qiskit-ibm.json in the same folder.

A quantum circuit for Listing 7-1 is shown in Figure 7-4. The circuit uses 3 qubits to generate an 8-bit random number between 0 and 255.

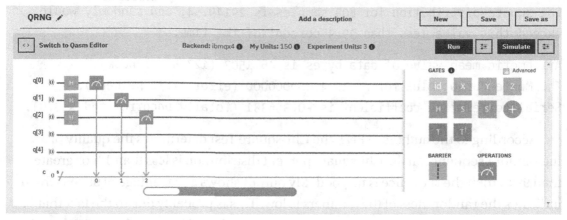

Figure 7-4. *Q Experience circuit for Listing 7-1*

Let's gather some data from multiple runs and put the results to the test.

Putting Randomness Results to the Test

Linux provides a neat program called *ent* (short for entropy) which is called a pseudorandom number sequence test program.[1] We can use this command to test the numbers generated in the previous section.

Tip   Windows users – a Windows32 binary is available for download from the ENT project site. A binary is also included in the source for this chapter under Workspace\Ch07\ent.exe. Mac users should be able to run the Linux binary from the command line.

Thus, generate 100 random 8-bit numbers using Listing 7-1, then run *ent*, to test your sequence with the command *ent [infile]* as shown in the next paragraph.

```
C:\Workspace\Ch07> python p7-1-qrng.py > qrnd-stdout.txt
C:\Workspace\Ch07>ent qrnd-stdout.txt
Entropy = 3.122803 bits per byte.
Optimum compression would reduce the size of this 805 byte file by 60 percent.
```

---

[1] ENT – A Pseudorandom Number Sequence Test Program available at www.fourmilab.ch/random/

Chi square distribution for 805 samples is 29149.54, and randomly would
exceed this value less than 99.9 percent of the time.

Arithmetic mean value of data bytes is 46.1503 (127.5 = random).
The Monte Carlo value for Pi is 4.000000000 (error 27.32 percent).
Serial correlation coefficient is -0.356331 (totally uncorrelated = 0.0).

According to the authors of ENT, the Chi-square Test determines the quality of
the random sequence. If the Chi-square percent distribution is less than 1% or greater
than 99%, then the sequence is no good. My output shows a percentage of 99.9% which
indicates the randomness of the numbers is low. This is probably due to the fact that I
used the remote simulator. This simulator is probably based in the default UNIX random
number generator (a poor-quality generator). See if your sequence does any better.
Table 7-1 shows the results from various deterministic and quantum sources head to
head provided by the developers of ENT.

Table 7-1. *Randomness test results from various sources gathered by ENT[1]*

Source	Chi-square percentage
UNIX rand()	99.9% for 500000 samples (bad)
Improved UNIX generator by Park and Miller	97.53% for 500000 samples (better)
HotBits: random numbers, generated by radioactive decay[2]	40.98% for 500000 samples (the best)

Table 7-1 shows that UNIX rand() shouldn't be trusted for random number
generation. If you need lots of truly random numbers (to generate encryption keys,
for example), use a quantum source such as HotBits. All in all, the purpose of this
section has been to get your feet wet with a simple quantum circuit for random number
generation. The next section takes things to the next level with the bizarre quantum data
transfer protocol dubbed super dense coding.

---

[2] HotBits: Genuine random numbers, generated by radioactive decay available online at
www.fourmilab.ch/hotbits/

Super Dense Coding

Super dense coding (SDC) is a data transfer protocol that demonstrates the remarkable information processing capabilities of a quantum system. Formally, SDC is a simple procedure that allows for transferring two classical bits of information to another party using a single qubit. The protocol is illustrated in Figure 7-5.

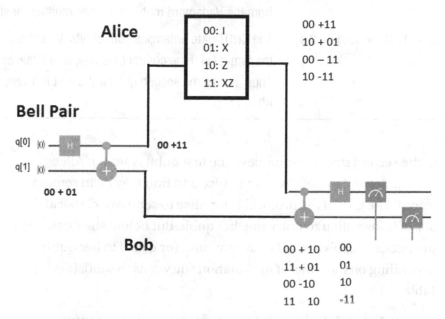

Figure 7-5. *Super dense coding protocol*

1. The process starts with a third party (Eve) generating what is called a Bell Pair. Eve starts with 2 qubits in the basis state |0>. She applies a Hadamard gate to the first qubit to create superposition. It then applies a CNOT gate using the first qubit as the control (dot) and the second as the target (+). This results in the states shown in Table 7-2.

Table 7-2. *Bell Pair states*

Gate	Outcome states	Details
H	$\lvert 00 \rangle \to \lvert 00 \rangle + \lvert 10 \rangle$	When the H gate is applied to the first qubit, it enters superposition; thus, we get the states 00 + 10 where the second qubit remains as 0. Note that the square root (2) from the Hadamard matrix has been omitted for simplicity
CNOT	$\lvert 00 \rangle + \lvert 10 \rangle \to \lvert 00 \rangle + \lvert 11 \rangle$	The CNOT gate entangles both qubits. In particular, it flips the target (+) if the control (.) is one, else it leaves intact. Thus, we flip the second qubit if the first is 1 resulting in 00 + 11

2. In the second step of the process, the first qubit is sent to Alice and the second to Bob. Note that Alice and Bob may be in remote places. The goal of the protocol is for Alice to send two classical bits of information to Bob using her qubit. But before she does, she needs to apply a set of quantum rules (or gates) to her qubit depending on the 2 bits of information she wants to send. (See Table 7-3.)

Table 7-3. *Encoding rules for super dense coding*

Rules	Outcome states
00: I (Identity gate)	I(00+11) = 00 + 11
01: X	X(00+11) = 10 + 01
10: Z	Z(00+11) = 00 − 11
11: ZX	ZX(00+11) = 10 − 11

3. Thus, if Alice sends a 00, she does nothing to her qubit (applies the identity gate). If she sends a 01, then she applies the X gate (or bit-flip). For a 10 she applies the Z gate. Note that the Z gate flips the sign (phase) of the qubit if the qubit is 1. Thus, $Z\lvert 0 \rangle = \lvert 0 \rangle, Z\lvert 1 \rangle = -\lvert 1 \rangle$. Finally, if she sends 11, then she applies gates XZ to her qubit. Alice then sends her qubit to Bob for the final step in the process.

4. Bob receives Alice's qubit (qubit-0) and uses his qubit to reverse
    the process of the Bell state created by Eve. That is, he applies the
    CNOT gate to the first qubit followed by the Hadamard gate (H)
    and finally performs a measurement in both qubits to extract the 2
    classical bits encoded in Alice's qubit (see Table 7-4).

Table 7-4. *Qubit states after recovery*

Gate	Outcome states	Details
CNOT	00 +10 11 + 01 00 -10 11 -10	We start with Alice's' states from step 2: 00 + 11 10 + 01 00 − 11 10 - 11 The CNOT gate flips the second qubit if the first is 1 resulting in the states in column #2
H	00 01 10 -11	Applying the Hadamard to the first qubit in the last row results in the outcomes in column #2. When Bob performs measurements in the computational basis states, he ends up with four possible outcomes with probability 1 each. These outcomes match what Alice meant to send in step 2 column 1. Note that the last outcome has a negative sign. Nevertheless, because the probability is calculated as the amplitude squared, the -1 becomes 1 which is correct

Let's put all this together in a circuit within the IBM Quantum composer.

Circuit for composer

Figure 7-6 shows the super dense coding circuit as well as the state vector probability within the composer:

- The circuit begins by creating a Bell Pair, that is, it puts qubit[0] in
  superposition (using the Hadamard gate) and then entangles it with
  qubit[1] via the CNOT gate.

- The next two gates represent Alice's encoding rules. Remember that she applies the identity (nothing) to encode bits 00, X to encode 01, Z to encode 10, and ZX to encode 11. In this particular case, the encoded bits are 11. This is shown on the left of the barrier symbol in Figure 7-6. Note that the barrier will block execution until all gates are consumed by both qubits.

- To the right side of the barrier symbol, there is Bob's protocol. He does the reverse operation as Alice's. He applies the CNOT gate and then a Hadamard gate on the qubits. Finally, a measurement is performed on both qubits to extract the two encoded classical bits.

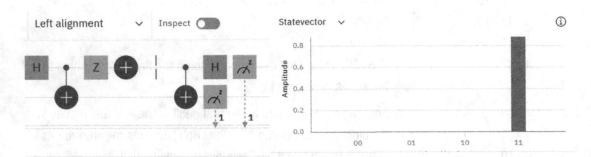

Figure 7-6. *Super dense circuit for the composer*

Run the preceding circuit in the simulator, and the result should be a bar graph with the probability for outcome 11 very close or equal to 1. This result should match the result obtained in the next section using a Python script.

Running in Python

Listing 7-2 shows the equivalent Python script for the circuit in Figure 7-6:

- Lines 17-19 create two qubits and two classical registers to hold the outcomes.

- Next the *superdense* circuit is created with the entangled Ball Pair (lines 22-14).

- Alice encodes 11 by applying the ZX gates. Optionally, comment any of these statements to encode a different pair, and then make sure the result matches Alice's encoding scheme (lines 32-35).

- Bob reverses Alice's operation and measures the qubits (lines 38-41).

- Finally, the circuit gets executed in the remote simulator (ibmq_
  qasm_simulator), and the results are displayed using Python's
  excellent plotting support.

Listing 7-2. Super dense coding Python script

```python
import sys,time,math

# Importing QISKit
from qiskit import *

# Import basic plotting tools
from qiskit.tools.visualization import plot_histogram

def main():

  # Creating registers
  q = QuantumRegister(2, "q")
  c = ClassicalRegister(2, "c")

  # Quantum circuit to make the shared entangled state
  superdense = QuantumCircuit(q, c, name="superdense")
  superdense.h(q[0])
  superdense.cx(q[0], q[1])

  # For 00, do nothing
  # For 10, apply X
  # superdense.x(q[0])
  # For 01, apply Z
  # superdense.z(q[0])

  # Alice: For 11, apply ZX
  superdense.z(q[0])
  superdense.x(q[0])
  superdense.barrier()

  # Bob
  superdense.cx(q[0], q[1])
  superdense.h(q[0])
```

```
superdense.measure(q[0], c[0])
superdense.measure(q[1], c[1])

superdense.draw(filename='superdense-circ.png')
circuits = [superdense]
print(superdense.qasm())

backend = Aer.get_backend("qasm_simulator")
result = execute(circuits, backend=backend).result()

print("Counts:" + str(result.get_counts("superdense")))
plot_histogram(result.get_counts(), filename="superdense-res.png")

##########################################
# main
if __name__ == '__main__':
  start_time = time.time()
  main()
  print("--- %s seconds ---" % (time.time() - start_time))
```

Let's look at the results of a single run of script 7-2 in the next section.

Looking at the Results

The standard output of a run of Listing 7-2 is shown in the next paragraph:

```
C:\python36-64\python.exe p07-superdensecoding.py
OPENQASM 2.0;
include "qelib1.inc";
qreg q[2];
creg c[2];
h q[0];
cx q[0],q[1];
z q[0];
x q[0];
barrier q[0],q[1];
cx q[0],q[1];
h q[0];
```

```
measure q[0] -> c[0];
measure q[1] -> c[1];

Counts:{'11': 1024}
--- 167.52969431877136 seconds ---
```

The script dumps the assembly code of the circuit as well as the counts for the
outcome: {'11': 1024} plus the execution time. The result count is used to calculate the
probability of the outcome by dividing the number of shots (1024) by the outcome
count (1024). Thus, the probability is 1 for outcome 11, as shown in the plot run as the
final step in Listing 7-2 (see Figure 7-7). Note that when executed in the simulator, the
probability will always be 1, that is, counts = shots. However, if you run in a real quantum
device, because of noise and environmental error, the number of counts should be less
than 1024 resulting in a probability less than 1.

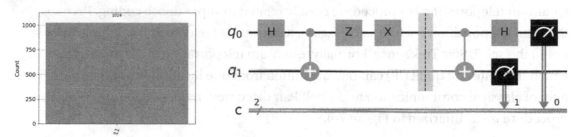

Figure 7-7. *Super dense coding result and circuit plots*

Thus, super dense coding provides the means to encode two classical bits in a single
qubit. Note that it is worth mentioning that quantum computation states that it is not
possible to store more than a single classical bit per qubit which seems to contradict
what has been shown in this protocol. As a matter of fact there is no contradiction. The
protocol works because Alice's and Bob's qubits are entangled via a Bell Pair. This allows
for sending two classical bits in Alice's entangled qubit. All in all, you can store at most
2 classical bits per qubit provided that your qubit is entangled to another via a Bell Pair.
Note that we are assuming that the Bell Pair can be transferred somehow between Alice
and Bob when in reality this is a difficult task and subject to extensive research.

In general terms, this protocol could be interpreted as a set of modularized
abstractions: a Bell Pair Generator module to create two entangled qubits, followed by
an information encoder module that applies Alice's rules to encode the 2 classical bits
of information. Finally, a decoder module extracts the classical bits from the qubits
provided by the Bell Pair as well as the encoder module (sort of a quantum

zip/unzip tool if you will). Super dense coding provides a high level picture for quantum information processing and will help you understand the next item in this chapter: quantum teleportation.

Tip   This simple protocol was developed in 1992 by physicist Charles Bennett almost 70 years after the discovery of quantum mechanics. Despite its relative simplicity, it is not an obvious procedure, and remarkable things can be learned by studying it in detail.

Quantum Teleportation

Quantum teleportation is a procedure closely related to super dense coding. Perhaps, the term teleportation is a little extravagant, as we are not teleporting anything, at least not in the sci-fi/Star Trek sense. Formally, quantum teleportation is the process by which the state of a qubit (Ψ) can be transmitted from one location to another, with the help of classical communication and a Bell Pair discussed in the previous section. The procedure is summarized in Figure 7-8.

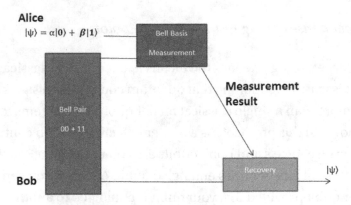

Figure 7-8.   *Quantum teleportation workflow*

1. Alice and Bob start by sharing a Bell Pair of entangled qubits. One goes to Alice and the other goes to Bob at separate remote locations. Imagine that the Bell Pair is prepared by a third party (Eve).

2. Alice prepares her qubit to be teleported in state $|\Psi\rangle = \alpha|0\rangle + \beta|1\rangle$.
    She then performs a Bell basis measurement of her qubit and the
    entangled qubit from the Bell Pair provided by Eve. Alice then
    sends the measurement result by classical means to Bob.

3. At this point, there is a posterior state for Bob's qubit as a
    function of the measurement performed by Alice. This is the
    key to understanding the procedure; remember that both share
    an entangled qubit. Thus, we'll see how Bob, by applying the
    appropriate quantum gate, can recover the original state Ψ
    created by Alice.

Let's figure this out by looking at Bob's posterior state at the moment of Alice's
measurement before the recovery operation. To do this, we write the joined states of the
three qubits involved in the process. Note that the ket notation is ignored for simplicity.
Thus, given Alice's state $|\Psi\rangle = \alpha|0\rangle + \beta|1\rangle$, if we combine it with the shared entangled
qubit from the Bell Pair provided by Eve, we get:

$$(\alpha 0 + \beta 1)\ (00 + 11) = \alpha 000 + \alpha 011 + \beta 100 + \beta 111 \qquad (1)$$

Now we need to write the state of the first two qubits using the Bell basis states

$$B0 = 00 + 11 \qquad 00 = B0 - B1$$
$$B1 = 10 + 01 \qquad 01 = B1 - B3$$
$$B2 = 00 - 11 \qquad 10 = B1 - B3$$
$$B3 = 10 - 01 \qquad 11 = B0 - B2$$

Expression (1) becomes:

$$(\alpha 0 + \beta 1)\ (00 + 11) = B0\ (\alpha 0 + \beta 1) + B1\ (\alpha 1 + \beta 0) + B2\ (\alpha 0 - \beta 1) +$$
$$B3\ (-\alpha 1 + \beta 0) \qquad (2)$$

Expression (2) shows the states for the three qubits after Alice performs her
measurement. Bob knows how to recover Alice's Ψ by looking at the posterior state of the
qubits in expression 2 (the states within the parenthesis). This is shown more clearly in
Table 7-5.

Table 7-5. *Quantum teleportation recovery*

Bell state	Posterior state	Bob's recovery operation
B0	$\alpha 0 + \beta 1$	Ψ
B1	$\alpha 1 + \beta 0$	$X\Psi$
B2	$\alpha 0 - \beta 1$	$Z\Psi$
B3	$-\alpha 1 + \beta 0$	$ZX\Psi$

All in all, the quantum teleportation protocol provides the means to recover the state Ψ of any qubit by sharing an entangled Bell Pair between two remote parties, hence the name teleportation. Now let's build a circuit for this protocol, run it in the simulator, and finally look at the results.

Circuit for composer

Figure 7-9 shows the composer circuit as well as the execution results (simulator only – no real device at this time) for the quantum teleportation protocol:

- The gates left of the barrier symbol (the dotted line) represent the Bell Pair prepared by the third party (Eve): qubits 1 and 2.

- Alice prepares her qubit (0) to a given state Ψ. The actual value of Ψ is irrelevant as it will be recovered by Bob at the final stage of the process. Alice receives qubit[1] from Eve; qubit[2] goes to Bob.

- Alice performs a measurement on her qubits [0,1] (shown to the right of the dotted line) and sends the results by classical means to Bob.

- Bob applies the recovery rules to his qubit (2) mentioned in the previous section depending on the outcomes sent by Alice. Finally, after a measurement of qubit[2], Bob recovers the state Ψ originally created by Alice. All this is made possible by the fact that Alice and Bob share an entangled pair of qubits which makes the whole thing work.

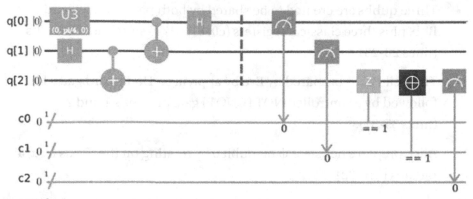

Teleportation

Device: Simulator

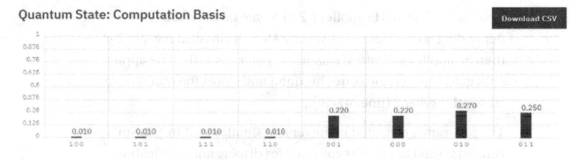

Figure 7-9. Quantum teleportation circuit for the composer

Of course, the execution results in Figure 7-9 need to be massaged to verify that Bob's
Ψ matches Alice's. The best way to do this is to use a Python script. In the next section,
we'll run the same circuit remotely and look at the results to verify the protocol works.

Running in Python

In this section we use Python to run the quantum teleportation protocol remotely in
the simulator. Note that, at this time, quantum teleportation cannot be run in a **real
quantum device** on IBM Q Experience. This is due to the fact the hardware does not
support the rotation gate required by Alice to create her state Ψ. Thus, we'll use the
remote simulator instead – a local Python simulator will be fine too. Listing 7-3 shows
the protocol in action. In particular

- Three qubits are created to be shared by both parties: Alice and Bob, plus three classical registers (c0, c1, c2) to store Alice's results (lines 20-23).

- The Bell Pair is prepared by Eve by applying a Hadamard gate (H) followed by a controlled NOT (CNOT) gate in qubits 1 and 2 (lines 35-37).

- Alice prepares her state ψ on qubit 0 by rotating on the Y axis by $\pi/4$ radians (line 32).

- Alice now entangles her qubit[0] with the Bell Pair qubit given to her, qubit[1], to entangle them. She then performs a measurement in both and stores the outcomes in classical registers 0, 1 (lines 35-41).

- Now its Bob's turn: He applies a Z or X gate on his qubit (2) depending on the outcomes sent by Alice: If classical register 0 is 1, then he applies a Z gate. If classical register 1 is 1, then he applies an X gate. Then he measures his qubit and stores the outcome in classical register 2 (lines 47-50).

- The program is executed in the remote simulator (ibmq_qasm_ simulator) and the results collected for display and verification (lines 58-79).

Tip   The source for this program is included in the book source under Workspace\ Ch07\p07-teleport.py.

Listing 7-3. Python script for quantum teleportation

```
import sys,time,math
import numpy as np

# Importing QISKit
from qiskit import *

# Import basic plotting tools
from qiskit.tools.visualization import plot_histogram
```

```python
def main():

  # Creating registers
  q = QuantumRegister(3, 'q')
  c = ClassicalRegister(3,'c')

  # Quantum circuit to make the shared entangled state (Bell Pair)
  teleport = QuantumCircuit(q, c, name='teleport')
  teleport.h(q[1])
  teleport.cx(q[1], q[2])

  # Alice prepares her quantum state to be teleported,
  # psi = a|0> + b|1> where a = cos(theta/2), b = sin (theta/2),
    theta = pi/4
  teleport.ry(np.pi/4,q[0])

  # Alice applies CNOT to her two quantum states followed by H, to
    entangle them
  teleport.cx(q[0], q[1])
  teleport.h(q[0])
  teleport.barrier()

  # Alice measures her two quantum states:
  teleport.measure(q[0], c[0])
  teleport.measure(q[1], c[1])

  # dump qasm. Note: cannot dump after c_if
  circuits = [teleport]
  print(circuits[0].qasm())

  ##### BOB Depending on the results applies X or Z, or both, to his state
  teleport.z(q[2]).c_if(c[0], 1)
  teleport.x(q[2]).c_if(c[1], 1)
  teleport.measure(q[2], c[2])

  # Execute in the simulator
  backend = Aer.get_backend("qasm_simulator")
  result = execute(circuits, backend=backend).result()

  print("Counts:" + str(result.get_counts("teleport")))
```

```
# RESULTS
# Alice's measurement:
data = result.get_counts('teleport')
print (data)

alice = {}
alice['00'] = data['000'] + data['100']
alice['10'] = data['010'] + data['110']
alice['01'] = data['001'] + data['101']
alice['11'] = data['011'] + data['111']
plot_histogram(alice, filename='alice.png')

#BOB
bob = {}
bob['0'] = data['000'] + data['010'] +  data['001'] + data['011']
bob['1'] = data['100'] + data['110'] +  data['101'] + data['111']
plot_histogram(bob,  filename='bob.png')
##########################################
# main
if __name__ ==  '__main__':
    start_time = time.time()
    main()
    print("--- %s seconds ---" % (time.time() - start_time))
```

To verify the results, the outcome counts returned by the simulator must be gathered for Alice and Bob. A plot of the results is the best way to verify that Alice's state Ψ has been recovered by Bob. Here is a sample of what the simulator returns:

```
{'100': 37, '101': 45, '111': 43, '011': 215, '001': 200, '000': 206,
'010': 230, '110': 48}
```

In this JSON string the left side is the outcome(s) of the three qubits in reverse order. For example, in the first outcome 1 0 0: B(1) A(0) A(0) for Alice = A and Bob = B. To the right is the count obtained for that specific outcome. Remember that the probability of this outcome (used for graphing purposes) is calculated by dividing by the total number of shots (1024), thus:

```
P(1 0 0) = 37/1024 = 0.036
```

The histogram plots for the results for Alice and Bob from the execution of Listing 7-3 are shown in Figure 7-10.

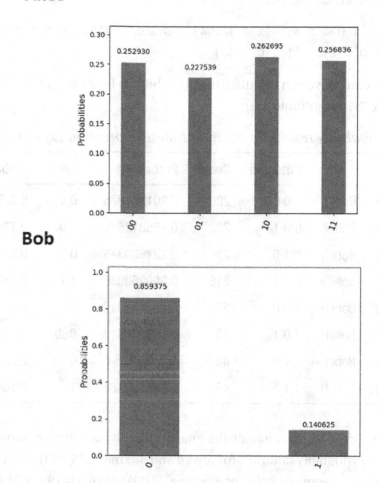

Figure 7-10. *Probability results for Alice and Bob measurements*

So what does this all mean? And how do we know that the state Ψ has been recovered by Bob? Let's look at these results in more detail.

Looking at the Results

To interpret these results, first let's see how the probabilities are calculated from the counts returned from Listing 7-3:

```
{'1 0 0': 37, '1 0 1': 45, '1 1 1': 43, '0 1 1': 215, '0 0 1': 200, '0 0
0': 206, '0 1 0': 230, '1 1 0': 48}
```

Using these counts we can calculate the probabilities for Alice's and Bob's outcomes shown in Figure 7-10 (see Table 7-6).

Table 7-6. *Probability results for the quantum teleportation experiment*

Row			Outcome	Count	Probability	Alice	Probability sum
0	Alice(00)	Bob(0)	0 0 0	206	0.201171875	0 0	0.237304688
1	Alice(01)	Bob(0)	0 0 1	200	0.1953125	1 0	0.239257813
2	Alice(10)	Bob(0)	0 1 0	230	0.224609375	0 1	0.271484375
3	Alice(11)	Bob(0)	0 1 1	215	0.209960938	1 1	0.251953125
4	Alice(00)	Bob(1)	1 0 0	37	0.036132813		
5	Alice(01)	Bob(1)	1 0 1	45	0.043945313	**Bob**	
6	Alice(10)	Bob(1)	1 1 0	48	0.046875	0	0.831054688
7	Alice(11)	Bob(1)	1 1 1	43	0.041992188	1	0.168945313

As shown in Table 7-6, to calculate the total probability of Alice's outcome 00, we need to sum the probability columns for rows 0 and 4. That is, P(A00) = 0.201 + 0.036 = 0.237. The same rules apply to Bob. For example, P(B0) = 0.20 + 0.19 + 0.22 + 0.20 = 0.83 (add probability columns for rows 0-3) This is shown on the right side for all outcomes of Alice and Bob. This is how the script in Listing 7-3 massages the data before plotting the results shown in Figure 7-10. But what does this mean, and how do we know that Bob has recovered Alice's Ψ? Let's look at Bob's total probability for his qubit:

Bob

0   0.20 + 0.19 + 0.22 + 0.20 = 0.83

1   0.036 + 0.043 + 0.046 + 0.041 = 0.168

Quantum mechanics says that the probability of Ψ is given by $P(\Psi) = |\Psi|^2$. That is, the probability density is the modulus squared of Ψ. Now remember that Alice prepared Ψ as:

$$\Psi = RY(\theta)\, Where\, \theta = \frac{\pi}{4}$$

That is, Alice applied a π/4 rotation over the Y-axis on her qubit. To see this more clearly, let's visualize the state Ψ using geometry (see Figure 7-11).

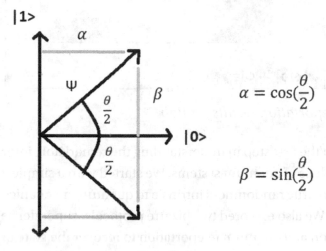

Figure 7-11. *Superimposed state for Alice's Ψ*

Remember that the superimposed state Ψ is described in terms of the complex coefficients α and β as:

$$\Psi = \alpha|0\rangle + \beta|1\rangle$$

$$Probability\,|0\rangle = |\alpha|^2, Probability\,|1\rangle = |\beta|^2$$

But from Figure 7-11 we can represent the coefficients as $\alpha = \cos(\theta/2)$ and $\beta = \sin(\theta/2)$. Thus finally, if θ= π/4 then:

$$Probability\,(\alpha) = |\cos(\pi/8)|^2 = 0.85$$
$$Probability\,(\beta) = |\sin(\pi/8)|^2 = 0.14$$

This matches Bob's results from the plot created by the teleportation script 7-3 (see Figure 7-12). Great success!

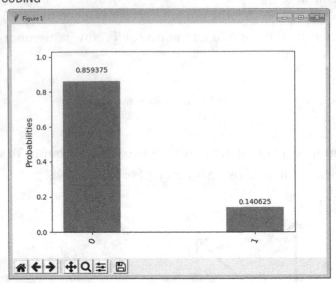

Figure 7-12. *Teleportation results for Bob*

You have taken the first step in understanding the remarkable information
processing capabilities of quantum systems. We started with a simple procedure to
exploit the source of true randomness intrinsic to quantum mechanics to generate
random numbers. We also explored two bizarre protocols: super dense coding to encode
classical information and quantum teleportation to recover the state of a qubit by a
remote party. These protocols have been described using circuits for IBM Quantum
as well as Python scripts for remote execution in a simulator or real quantum device.
Results have been collected and explained to understand what goes on behind the
scenes. The next chapter explores the lighter side of quantum computing, by having fun
creating a simple game using quantum gates: A break before we get to heavy stuff in later
chapters.

Exercises

7.1 Predict the output counts for the given snippet.

```
q = QuantumRegister(2,'q')
c = ClassicalRegister(2,'c')
qc = QuantumCircuit(q,c)
qc.h(0)
```

```
qc.h(1)
qc.measure([0,1],[0,1])
backend = BasicAer.get_backend('qasm_simulator')
job = execute(qc, backend, shots=100)
counts = job.result().get_counts()
```

a. {'11': 50, '00': 50}

b. {'00': 100}

c. {'11': 30, '01': 27, '10': 22, '00': 21}

d. {'11': 100}

7.2 What is the output if the following GHZ-circuit is simulated 100 times? Tip:

$$|W\rangle = \frac{1}{\sqrt{2}}(|000\rangle + |111\rangle)$$

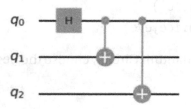

```
{'101': 50, '010': 50}
{'000': 51, '111': 49}
{'111': 5, '000': 50, '101': 45}
```

7.3 What is the result of CNOT(1/sqrt(2) (|10> + |11>)) if qubit-1 is the control and the qubit-0 is the target? (Use little endian bit ordering.)

```
1/sqrt(2) (|10> + |10>)
1/sqrt(2) (|11> + |10>)
|11>
```

7.4 Which quantum circuits will produce a Bell state (maximum entangled state)? (Select 3)

```
bell = QuantumCircuit(2)
bell.h(0)   #A
bell.cx(0,1)
```

```
bell.h(o) # B
bell.x(1)
bell.cx(0,1)
bell.h(o) # C
bell.h(1)
bell.cx(0,1)
bell.h(o) # D
bell.x(1)
bell.cx(0,1)
bell.z(1)
```

7.5 In the QuantumCircuit below, how many Qubits are there?

```
bob   = QuantumRegister(8,'b')
alice = ClassicalRegister(2,'a')
eve   = QuantumRegister(4,'e')
qc = QuantumCircuit(bob,alice,eve)
```

7.6 Given the following Quantum Circuit, choose the best option to run it 2000 times in the qasm_simulator.

```
qc= QuantumCircuit(2)
qc.h(0)
qc.cx(0,1)
sim = Aer.get_backend('qasm_similator')
a) execute (qc, sim, shots=2000)
b) execute (qc, sim, shot=2000)
c) execute (qc, sim, repeat=2000)
d) execute (qc, sim, repeats=2000)
```

7.7 In the circuit given below using the unitary simulator as the backend, choose the _missing_element_ from the options.

```
qc = QuantumCircuit(1)
qc.h(0)
backend_unitary = BasicAer.get_backend('unitary_simulator')
result = execute(qc,backend_unitary).result()._missing_element_
```

a) get_unitary()

b) get_unitary_matrix()

c) get_unitary_result()

d) get_unitary_simulator()

7.8 To return the individual list of measurement outcomes for each individual shot (provided the backend supports it), which argument needs to be set to true in the execute function?

memory, memory_shots, shots, single_shot

7.9 Which one of the below statements is invalid when drawing the quantum circuit?

a) qc.draw(output='mpl1')

b) qc.draw(output='text')

c) qc.draw(output='latex')

d) qc.draw(output='png')

7.10 Which code snippet would execute a circuit given these parameters?

- Use the QASM simulator

- Use a coupling map that connects three qubits linearly

- Run the circuit 100 times

a) sim = Aer.get_backend('ibm_similator)
execute (qc, loop=100, copulping_map=[[0,1],[0,2])

b) sim = Aer.get_backend('qasm_similator)
execute (qc, sim, shots= 100, copulping_map=[[0,1],[1,2])

c) sim = Aer.get_backend('qasm_similator)
execute (qc, sim, repeat= 100, copulping_map=[[0,1],[1,2])

## CHAPTER 8

Game Theory: With Quantum Mechanics, Odds Are Always in Your Favor

This chapter explores two game puzzles that show the remarkable power of quantum algorithms over their classical counterparts:

- The counterfeit coin puzzle: It is a classical balance puzzle proposed by mathematician E.D. Schell in 1945. It is about balancing coins to determine which holds a different value (counterfeit) using a balance scale and a limited number of tries.

- The Mermin-Peres Magic Square Game: This is an example of quantum pseudo-telepathy or the ability of players to achieve outcomes that would only be possible if they mysteriously communicate during the game.

In both cases, quantum computation gives quasi-magical abilities to the players, just as if they were cheating all along. Let's see how.

© Vladimir Silva 2024

V. Silva, *Quantum Computing by Practice*, https://doi.org/10.1007/978-1-4842-9991-3_8

Counterfeit Coin Puzzle

In this puzzle, the player has eight coins and a beam balance. One of the coins is fake and thus underweight. The goal of the game is to figure out which coin is fake by using the balance only twice. Can you figure out how? Let's run through the solution shown in Figure 8-1.

1. Given 8 coins put coins (1-3) on the left side of the balance, 4-6 on the right side. Leave the last 2 coins 7 and 8 on the side, and weigh.

2. If the balance leans right, the counterfeit is among 1-3 (left). Remember that the fake coin is lighter. Thus, take out the last coin from the left (3) and weigh it again (for the second time).

   - If the beam leans right, the counterfeit is 1. Stop.

   - If the beam leans left, the counterfeit is 2. Stop.

3. If the balance leans left, the counterfeit is within 4-6. Take out the last coin (6) and weigh it again.

   - If the beam leans right, the counterfeit is 5. Stop.

   - If the beam leans left, the counterfeit is 6. Stop.

4. If the beam is balanced, the counterfeit is either 7 or 8. Put 7 and 8 in the balance and weigh again.

   - If the beam leans right, the counterfeit is 7. Stop.

   - If the beam leans left, the counterfeit is 8. Stop.

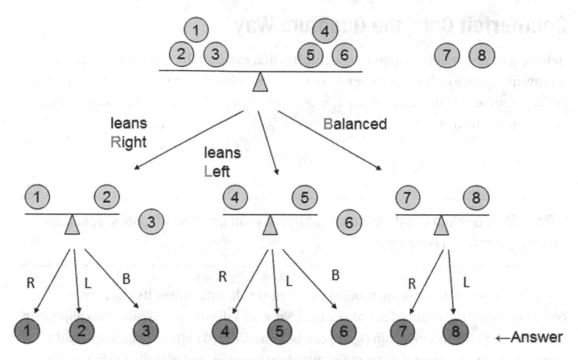

Figure 8-1. *Counterfeit puzzle solution for eight coins*

From the procedure in the previous section, a classical algorithm can be implemented independent of the total number of coins N and the number of counterfeit coins k. In general terms, the time complexity of the generalized counterfeit coin puzzle is given by:

$$O(k\log(N/k))$$

Tip   It has been proven that the minimal number of tries required to find a single counterfeit coin using the balance beam in a classical computer is two.

Counterfeit Coin, the Quantum Way

Believe it or not, there is a quantum algorithm that can find the counterfeit using a quantum balance only once, independent of the number of coins N! In general terms, for any number of counterfeit coins k, independent of N, the time complexity of such algorithm is given by:

$$O\left(k^{1/4}\right)$$

Tip   The quantum counterfeit coin algorithm is an example of quartic speedup over its classical counterpart.

Thus Figure 8-2 shows the power of a quantum algorithm over its classical counterpart for the counterfeit coin puzzle. Now, let's dig deeper: A quantum algorithm to find a single counterfeit coin (k = 1) can be summarized in three stages: query the quantum beam balance, construct the quantum balance, and identify the false coin.

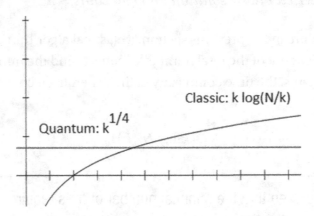

Figure 8-2. *Quantum vs. classical time complexities for the counterfeit coin puzzle*

Step 1: Query the Quantum Beam Balance

A quantum algorithm will query the beam balance in superposition. To do this, we use a binary query string to encode coins placed on the pans. For example, the query string 11101111 means all coins are on the beam except coin with index 3. The beam is balanced when no false coin is included, and tilted otherwise. The next table illustrates this.

N (# of coins)	F (index of false coin)	Query string	Result
8	3	11101111	balanced (0)
8	3	11111111	tilted (1)

The procedure can be summarized as follows:

1. Use two quantum registers to query the quantum balance, where the first register is for the query string and the second register is for the result.

2. Prepare the superposition of all binary query strings with an even number of 1s.

3. To obtain the superposition of states of an even number of 1s, perform a Hadamard transformation on the basis state |0>, and check if the Hamming weight of |x| is even. It can be shown that the Hamming weight of |x| is even if and only if $x1 \oplus x2 \oplus \ldots \oplus xN = 0$

Tip   The Hamming weight (hw) of a string is the number of symbols that are different from the zero-symbol of the alphabet used. For example: hw(11101) = 4, hw(11101000) − 4, hw(000000) − 0.

---

4. Finally, measure the second register, and if |0> is observed, then the first register is the superposition of all binary query strings we want. If we get |1>, then repeat the procedure until |0> is observed.

Note that each repetition is guaranteed to succeed with a probability of exactly half. Hence, after several repetitions, we should be able to obtain the desired superposition state. Listing 8-1 shows an implementation of a quantum program to query the beam balance with the corresponding graphical circuit shown in Figure 8-3.

Note   For illustration purposes the full counterfeit coin program has been broken in Listings 8-1 thru 8-3; to run the program, use the full listing available from the source at Workspace\Ch08\p_counterfeitcoin.py.

Listing 8-1. Script to query the quantum beam balance

```
# ------- Query the quantum beam balance
qr = QuantumRegister(numberOfCoins + 1, "qr")

# for recording the measurement on qr
cr = ClassicalRegister(numberOfCoins+1, "cr")

circuitName = "QueryStateCircuit"
circuit   = QuantumCircuit( qr, cr, name = circuitName)

N = numberOfCoins

#Create uniform superposition of all strings of length N
for i in range(N):
  circuit.h(qr[i])

#Perform XOR(x) by applying CNOT gates sequentially from qr[0]
#to qr[N-1] and storing the result to qr[N]
for i in range(N):
  circuit.cx(qr[i], qr[N])

# Measure qr[N] and store the result to cr[N].
# We continue if cr[N] is zero, or repeat otherwise
circuit.measure(qr[N], cr[N])

# Query the quantum beam balance
# if the value of cr[0]...cr[N] is all zero
# by preparing the Hadamard state of |1>, i.e., |0> - |1> at qr[N]
circuit.x(qr[N]).c_if(cr, 0)
circuit.h(qr[N]).c_if(cr, 0)
```

```
# we rewind the computation when cr[N] is not zero
for i in range(N):
  circuit.h(qr[i]).c_if(cr, 2**N)
```

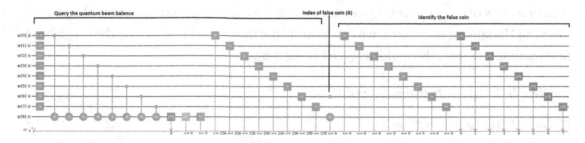

Figure 8-3. *Quantum circuit for the counterfeit coin puzzle with N = 8, k = 1, and fake at index 6*

Figure 8-3 shows a complete circuit for the counterfeit coin puzzle for 8 coins, 1 counterfeit at index 6. The circuit displays all the stages described here for the IBM Q Experience platform. The second stage in the algorithm is to construct the beam balance.

Step 2: Construct the Quantum Balance

In the previous section, we constructed the superposition of all binary query strings whose Hamming weights are even. In this step, we construct the quantum beam by setting the position of the false coin. Thus, given k the position of the false coin with regard to the binary string $|x1, x2, ..., xN>|0>$, the quantum beam balance returns:

$$|x1, x2, ... , xN> |0{\oplus}xk>$$

This is implemented with a CNOT gate with xk as the control and the second register as the target (see partial Listing 8-2).

Listing 8-2. Construct the quantum beam balance.

```
#----- Construct the quantum beam balance
k = indexOfFalseCoin

# Apply the quantum beam balance on the desired superposition state
#(marked by cr equal to zero)
circuit.cx(qr[k], qr[N]).c_if(cr, 0)
```

Step 3: Identify the False Coin

To identify the false coin after querying the balance, apply a Hadamard transformation on the binary query string. Assuming that we query the quantum beam balance with binary strings of even Hamming weight, then by measuring the computational basis after the Hadamard transform, we can identify the false coin because it is the one whose label is different from the majority (see Listing 8-3).

Listing 8-3. Identify the false coin.

```
# --- Identify the false coin
# Apply Hadamard transform on qr[0] ... qr[N-1]
for i in range(N):
  circuit.h(qr[i]).c_if(cr, 0)

# Measure qr[0] ... qr[N-1]
for i in range(N):
  circuit.measure(qr[i], cr[i])

results  = execute([circuit], backend=backend, shots=shots).result()
answer  = results.get_counts(circuitName)

print("Device " + str(backend) + " counts " + str(answer))

# Get the most common label
for key in answer.keys():
  normalFlag, _ = Counter(key[1:]).most_common(1)[0]
  for i in range(2,len(key)):
    if key[i] != normalFlag:
      print("False coin index is: ", len(key) - i - 1)
```

When the left-most bit is 0, the index of the false coin can be determined by finding the one whose values are different from others. For example, for N =8, and false index 6, the result should be 010111111 or 001000000. Note that because we use cr[N] to control the operation prior to and after the query to the balance, thus

- If the left-most bit is 0, then we succeed in identifying the false coin.

- If the left-most bit is 1, we failed to obtain the desired superposition and must repeat the process from the beginning. This may result in a different simulator count result from what I show in the following; also keep in mind that the program may fail to identify the false coin; thus you may need to run the script a few times. All in all, the execution parameters can be tuned in the script such as index of the false coin, number of qubits, and total number of coins.

Running the program against the remote IBM-Q simulator gives the result (Under book source Workspace\Ch07\p_counterfeitcoin.py). Note that I am using Windows in this instance:

```
c:\python36-64\python.exe p_counterfeitcoin.py
Device ibmq_qasm_simulator counts {'001000000': 1}
False coin index is:  2
```

If you don't have access to the book source and still want to play with this script, Listing 8-4 is a condensed version of the whole thing. Give it a try.

Listing 8-4. Counterfeit coin puzzle main container script

```
import sys
import matplotlib.pyplot as plt
import numpy as np

# useful math functions
from math import pi, cos, acos, sqrt
from collections import Counter

# importing the QISKit
from qiskit import *

# import basic plot tools
from qiskit.tools.visualization import plot_histogram

def main(M = 16, numberOfCoins = 8 , indexOfFalseCoin = 6
  , backend = None , shots = 1 ):
```

```
if numberOfCoins < 4 or numberOfCoins >= M:
   raise Exception("Please use numberOfCoins between 4 and ", M-1)
if indexOfFalseCoin < 0 or indexOfFalseCoin >= numberOfCoins:
   raise Exception("indexOfFalseCoin must be between 0 and ",
numberOfCoins-1)

# ------- Query the quantum beam balance
# Create registers
qr = QuantumRegister(numberOfCoins + 1, "qr")

# for recording the measurement on qr
cr = ClassicalRegister(numberOfCoins+1, "cr")

circuitName = "QueryStateCircuit"
circuit    = QuantumCircuit( qr, cr, name = circuitName)

N = numberOfCoins

#Create uniform superposition of all strings of length N
for i in range(N):
   circuit.h(qr[i])

#Perform XOR(x) by applying CNOT gates sequentially from qr[0]
#to qr[N-1] and storing the result to qr[N]
for i in range(N):
   circuit.cx(qr[i], qr[N])

# Measure qr[N] and store the result to cr[N].
# We continue if cr[N] is zero, or repeat otherwise
circuit.measure(qr[N], cr[N])

# we proceed to query the quantum beam balance
# if the value of cr[0]...cr[N] is all zero
# by preparing the Hadamard state of |1>, i.e., |0> - |1> at qr[N]
circuit.x(qr[N]).c_if(cr, 0)
circuit.h(qr[N]).c_if(cr, 0)

# we rewind the computation when cr[N] is not zero
for i in range(N):
```

```
    circuit.h(qr[i]).c_if(cr, 2**N)

  #----- Construct the quantum beam balance
  k = indexOfFalseCoin

  # Apply the quantum beam balance on the desired superposition
  # state (marked by cr equal to zero)
  circuit.cx(qr[k], qr[N]).c_if(cr, 0)

  # --- Identify the false coin
  # Apply Hadamard transform on qr[0] ... qr[N-1]
  for i in range(N):
    circuit.h(qr[i]).c_if(cr, 0)

  # Measure qr[0] ... qr[N-1]
  for i in range(N):
    circuit.measure(qr[i], cr[i])

  print(circuit.qasm())


  results    = execute([circuit], backend=backend, shots=shots).result()
  answer     = results.get_counts(circuitName)

  print("Device " + str(backend) + " counts " + str(answer))

  #plot_histogram(answer)

  for key in answer.keys():
    normalFlag, _ = Counter(key[1:]).most_common(1)[0] #get most
    common label
    for i in range(2,len(key)):
      if key[i] != normalFlag:
        print("False coin index is: ", len(key) - i - 1)

####################################################
# main
####################################################
if __name__ == '__main__':
  M = 8                    #Maximum qubits available
  numberOfCoins = 4     #Up to M-1, where M is the number of qubits
```

```
indexOfFalseCoin = 2 #This should be 0, 1, ..., numberOfCoins - 1,

# simulator
name    = "qasm_simulator"
backend = Aer.get_backend(name)

shots   = 1      # We perform a one-shot experiment
main(M, numberOfCoins, indexOfFalseCoin, backend, shots)
```

Generalization for Any Number of False Coins

The counterfeit coin puzzle has been generalized by any number of fake coins (k>1) by mathematicians Terhal and Smolin in 1998. Their implementation uses a Balance Oracle model (B-Oracle) such that

- Given an input of N bits $x = x1x2...xn \in \{0, 1\}^N$

- Construct a query string of N tri-bits such that $q = q1q2... qn \in \{0, 1, -1\}^N$ with the same number of 1s and -1s.

- The answer is 1 bit such that

$$\chi(x; q) = \begin{cases} 0 \text{ if } q1x1 + q2x2 + ...qnxn = 0 \,(balanced) \\ 1 \text{ otherwise} \qquad\qquad\qquad (tilted) \end{cases}$$

Tip   An oracle is the portion of an algorithm regarded as a black box. It is used to simplify circuits and provide complexity comparisons between quantum and classical algorithms. A good oracle should provide speed, generality, and feasibility.

An example of the B-Oracle in action is shown in Figure 8-4 for 2 fake coins: k = 2 and N= 6.

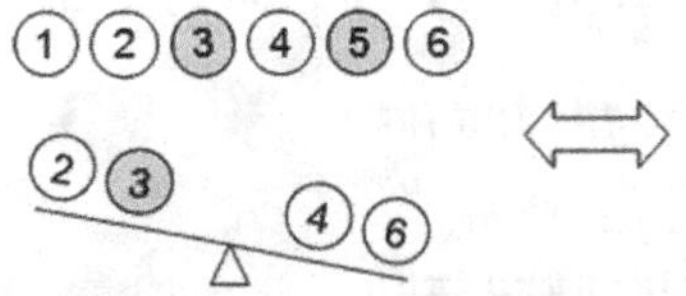

$$x = 001010$$
$$q = 0 - 1 - 1101$$
$$\chi(x; q) = 1 \ (\because \sum x_i q_i = -1)$$

Figure 8-4. *B-Oracle for N = 6 and k = 2*

All in all, the counterfeit puzzle is the quintessential example of quartic speedup of a quantum algorithm over its classical counterpart. In the next section, we look at another bizarre quasi-magical puzzle called the Mermin-Peres Magic Square.

Mermin-Peres Magic Square

This is another classic puzzle first proposed by physicists David Mermin and A. Peres as an example of quantum pseudo-telepathy or the ability of two players to have some supernatural communication to outside observers. Thanks to the magic of entanglement, this is possible. Let's take a closer look.

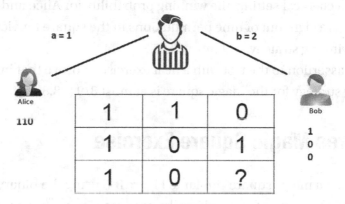

Figure 8-5. *Mermin-Peres Magic Square*

The game starts with two players Alice and Bob against a referee. The magic square is a 3x3 matrix with the following rules (see Figure 8-5):

- All entries are either 0 or 1 such that the sum of entries on each row is even, and the sum of each column is odd. The game is called the magic square because such a square is impossible, as shown in Figure 8-5; there is no valid combination where the sum of rows is even and the sum of columns is odd (try it yourself with pen and paper). This is due to the odd number of entries in the matrix.

- The referee sends an integer $a \in \{1,2,3\}$ to Alice and another $b \in \{1,2,3\}$ to Bob. Alice must reply with the a-th row of the square. Bob must reply with the b-th column.

- Alice and Bob win if the sum of Alice's entries is even, the sum of Bob's is odd, and their intersecting answer is the same. Otherwise, the referee wins.

- Prior to the start, Alice and Bob can strategize and share information. For example, they could decide to answer using the matrix in Figure 8-5. However, they are not allowed to communicate during the game.

For example, in the preceding matrix, if the referee sends a = 1 to Alice and b = 2 to Bob, Alice would reply with 110 (row 1) and Bob with 100 (column 2). The element in the intersection of the answers (row1-column2) is the same (1), so they win the game. It can be shown that in a classical setting, the winning probability for Alice and Bob is at most 8/9. That is, there are eight out of nine permutations in the square for victory. Therefore, Alice and Bob's winning strategy is at most **88.8%**.

Let's put this assertion to the test with a neat exercise to prove that indeed the classical winning strategy for the magic square is at most 8/9 (88.88%).

Mermin-Peres Magic Square Exercise

1. Construct a magic square similar to Figure 8-5 using the binary code (1,-1) instead of (1, 0) where the product of the row elements is 1 (even), and the product of the column elements is -1 (odd). Confirm that, in fact, this is not possible.

2. Create a permutation table for the referee values for a and b using the square in step 1 including

   - A permutation count number.

   - The values for a, b.

   - Alice and Bob's response.

   - The intersection of Alice and Bob's response. Remember that it must be equal for them to win.

   - The result of the game iteration: Win = W, Loose = L.

3. Finally, calculate the winning probability and prove that it is at most 8/9. **Note: answers are at the end of the section**.

Quantum Winning Strategy

Thanks to the power of quantum mechanics and the magic of entanglement, Alice and Bob can do much better. In fact, they can win the game 100% of the time. As if they were communicating telepathically, hence the term pseudo-telepathy. A quantum winning strategy was first proposed by Brassard and colleagues,[1] and it is divided into three stages:

- Shared entangled state: This is the key for Alice and Bob to win 100% of the time.

- Unitary transformations for Alice and Bob's inputs: These provide the responses to be sent back to the referee.

- Measure in the computational basis: The final stage to construct a final response.

Shared Entangled State

In the quantum winning strategy, Alice and Bob share the entangled state:

$$\Psi = \frac{1}{2}|0011\rangle - \frac{1}{2}|0110\rangle - \frac{1}{2}|1001\rangle + \frac{1}{2}|1100\rangle$$

A circuit implementation requires 2 qubits for Alice and 2 for Bob as shown in Figure 8-6.

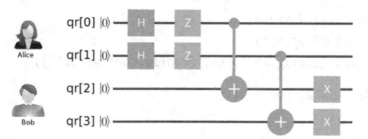

Figure 8-6. *Entangled state for the magic square*

---

[1] Brassard, Broadbent, and Tapp. Quantum Pseudo-Telepathy. pp 22, available online at https://arxiv.org/abs/quant-ph/0407221v3

- We know that the Hadamard maps the basis state
  $H|0\rangle \rightarrow \frac{1}{\sqrt{2}}(|0\rangle+|1\rangle)$. Thus, applying for the first two qubits

  yields: $\Psi = \frac{1}{2}|00\rangle + \frac{1}{2}|01\rangle + \frac{1}{2}|10\rangle + \frac{1}{2}|11\rangle$.

- Next, apply a Z gate to the first two qubits. Remember that
  Z leaves the 0 state unchanged and maps 1 to -1 (flipping
  the sign of the third term above). At this stage the state

  becomes: $\Psi = \frac{1}{2}|00\rangle + \frac{1}{2}|01\rangle - \frac{1}{2}|10\rangle + \frac{1}{2}|11\rangle$.

- Next, apply the CNOT gate to entangle qubits 0-2

  and 1-3: $\Psi = \frac{1}{2}|0000\rangle - \frac{1}{2}|0101\rangle - \frac{1}{2}|1010\rangle + \frac{1}{2}|1111\rangle$.

- Finally, flip the last 2 qubits with the X

  gate for $\Psi = \frac{1}{2}|0011\rangle - \frac{1}{2}|0110\rangle - \frac{1}{2}|1001\rangle + \frac{1}{2}|1100\rangle$.

The Python script to construct the entangled state is given in Listing 8-5.

Note   The magic square program has been broken down into listings 8-5, 8-6, and 8-7. You can run the full script from the chapter source at Workspace\Ch08\p_ magicsq.py.

Listing 8-5. Quantum winning strategy entangled state

```
# Create the entangled state
Q_program = QuantumProgram()
Q_program.set_api(Qconfig.APItoken, Qconfig.config["url"])

# 4 qubits (Alice = 2, Bob = 2)
N = 4

# Creating registers
qr = Q_program.create_quantum_register("qr", N)

# for recording the measurement on qr
cr = Q_program.create_classical_register("cr", N)
```

```
circuitName = "sharedEntangled"
sharedEntangled = Q_program.create_circuit(circuitName, [qr], [cr])

#Create uniform superposition of all strings of length 2
for i in range(2):
    sharedEntangled.h(qr[i])

#The amplitude is negative for an odd number of 1s
for i in range(2):
    sharedEntangled.z(qr[i])

#Copy the content of the first two qubits to the last two qubits
for i in range(2):
    sharedEntangled.cx(qr[i], qr[i+2])

#Flip the last two qubits
for i in range(2,4):
    sharedEntangled.x(qr[i])
```

With the shared entangled state, Alice and Bob can now start the game and receive inputs from the referee.

Unitary Transformations

Upon receiving their inputs $a \in \{1,2,3\}$ and $b \in \{1,2,3\}$, Alice and Bob apply the following unitary transformations: A1, A2, A3 for Alice and B1, B2, B3 for Bob to the shared entangled states:

$$A1 = \frac{1}{\sqrt{2}}\begin{bmatrix} i & 0 & 0 & 1 \\ 0 & -i & 1 & 0 \\ 0 & i & 1 & 0 \\ 1 & 0 & 0 & i \end{bmatrix}, A2 = \frac{1}{2}\begin{bmatrix} i & 1 & 1 & i \\ -i & 1 & -1 & i \\ i & 1 & 1 & -i \\ -i & 1 & -1 & -i \end{bmatrix}, A3 = \frac{1}{2}\begin{bmatrix} -1 & -1 & -1 & 1 \\ 1 & 1 & -1 & 1 \\ 1 & -1 & 1 & 1 \\ 1 & -1 & -1 & -1 \end{bmatrix}$$

$$B1 = \frac{1}{2}\begin{bmatrix} i & -i & 1 & 1 \\ -i & -i & 1 & -1 \\ 1 & 1 & -i & i \\ -i & i & 1 & 1 \end{bmatrix}, B2 = \frac{1}{2}\begin{bmatrix} -1 & i & 1 & i \\ 1 & i & 1 & -i \\ 1 & -i & 1 & i \\ -1 & -i & 1 & -i \end{bmatrix}, B3 = \frac{1}{\sqrt{2}}\begin{bmatrix} 1 & 0 & 0 & 1 \\ -1 & 0 & 0 & 1 \\ 0 & 1 & 1 & 0 \\ 0 & 1 & -1 & 0 \end{bmatrix}$$

Note   Remember that by applying the above transformations to their entangled states, Alice and Bob can construct the first 2 bits of their respective responses to the referee.

Listing 8-6 shows the unitary transformations for Alice and Bob with equivalent graphical circuits in Table 8-1.

Listing 8-6. Unitary transformations for Alice and Bob

```python
#------  circuits of Alice's and Bob's operations.
#we first define controlled-u gates required to assign phases
from math import pi

def ch(qProg, a, b):
    """ Controlled-Hadamard gate """
    qProg.h(b)
    qProg.sdg(b)
    qProg.cx(a, b)
    qProg.h(b)
    qProg.t(b)
    qProg.cx(a, b)
    qProg.t(b)
    qProg.h(b)
    qProg.s(b)
    qProg.x(b)
    qProg.s(a)
    return qProg

def cu1pi2(qProg, c, t):
    """ Controlled-u1(phi/2) gate """
    qProg.u(0,0,pi/4.0, c)
    qProg.cx(c, t)
    qProg.u(0,0,-pi/4.0, t)
    qProg.cx(c, t)
    qProg.u(0,0,pi/4.0, t)
    return qProg
```

```
def cu3pi2(qProg, c, t):
  """ Controlled-u3(pi/2, -pi/2, pi/2) gate """
  qProg.u(0,0,pi/2.0, t)
  qProg.cx(c, t)
  qProg.u(-pi/4.0, 0, 0, t)
  qProg.cx(c, t)
  qProg.u(pi/4.0, -pi/2.0, 0, t)
  return qProg

#-----------------------------------------------------------------------
# Define circuits used by Alice and Bob for each of their inputs: 1,2,3
# dictionary for Alice's operations/circuits
aliceCircuits = {}

# Quantum circuits for Alice 1, 2, 3
for idx in range(1, 4):
  circuitName = "Alice"+str(idx)
  aliceCircuits[circuitName]
    = QuantumCircuit (qr, cr, name = circuitName)
  theCircuit = aliceCircuits[circuitName]

  if idx == 1:
    #the circuit of A_1
    theCircuit.x(qr[1])
    theCircuit.cx(qr[1], qr[0])
    theCircuit = cu1pi2(theCircuit, qr[1], qr[0])
    theCircuit.x(qr[0])
    theCircuit.x(qr[1])
    theCircuit = cu1pi2(theCircuit, qr[0], qr[1])
    theCircuit.x(qr[0])
    theCircuit = cu1pi2(theCircuit, qr[0], qr[1])
    theCircuit = cu3pi2(theCircuit, qr[0], qr[1])
    theCircuit.x(qr[0])
    theCircuit = ch(theCircuit, qr[0], qr[1])
    theCircuit.x(qr[0])
    theCircuit.x(qr[1])
```

277

```
        theCircuit.cx(qr[1], qr[0])
        theCircuit.x(qr[1])

      elif idx == 2:
        theCircuit.x(qr[0])
        theCircuit.x(qr[1])
        theCircuit = cu1pi2(theCircuit, qr[0], qr[1])
        theCircuit.x(qr[0])
        theCircuit.x(qr[1])
        theCircuit = cu1pi2(theCircuit, qr[0], qr[1])
        theCircuit.x(qr[0])
        theCircuit.h(qr[0])
        theCircuit.h(qr[1])

      elif idx == 3:
        theCircuit.cz(qr[0], qr[1])
        theCircuit.swap(qr[0], qr[1]) # not supported in composer
        theCircuit.h(qr[0])
        theCircuit.h(qr[1])
        theCircuit.x(qr[0])
        theCircuit.x(qr[1])
        theCircuit.cz(qr[0], qr[1])
        theCircuit.x(qr[0])
        theCircuit.x(qr[1])

      #measure the first two qubits in the computational basis
      theCircuit.measure(qr[0], cr[0])
      theCircuit.measure(qr[1], cr[1])

# dictionary for Bob's operations/circuits
bobCircuits = {}

# Quantum circuits for Bob when receiving 1, 2, 3
for idx in range(1,4):
    circuitName = "Bob"+str(idx)
    bobCircuits[circuitName]
        = QuantumCircuit( qr, cr, name = circuitName)
    theCircuit = bobCircuits[circuitName]
```

```python
if idx == 1:
  theCircuit.x(qr[2])
  theCircuit.x(qr[3])
  theCircuit.cz(qr[2], qr[3])
  theCircuit.x(qr[3])
  theCircuit.u1(pi/2.0, qr[2])
  theCircuit.x(qr[2])
  theCircuit.z(qr[2])
  theCircuit.cx(qr[2], qr[3])
  theCircuit.cx(qr[3], qr[2])
  theCircuit.h(qr[2])
  theCircuit.h(qr[3])
  theCircuit.x(qr[3])
  theCircuit = cu1pi2(theCircuit, qr[2], qr[3])
  theCircuit.x(qr[2])
  theCircuit.cz(qr[2], qr[3])
  theCircuit.x(qr[2])
  theCircuit.x(qr[3])

elif idx == 2:
  theCircuit.x(qr[2])
  theCircuit.x(qr[3])
  theCircuit.cz(qr[2], qr[3])
  theCircuit.x(qr[3])
  theCircuit.u1(pi/2.0, qr[3])
  theCircuit.cx(qr[2], qr[3])
  theCircuit.h(qr[2])
  theCircuit.h(qr[3])

elif idx == 3:
  theCircuit.cx(qr[3], qr[2])
  theCircuit.x(qr[3])
  theCircuit.h(qr[3])

#measure the third and fourth qubits in the computational basis
theCircuit.measure(qr[2], cr[2])
theCircuit.measure(qr[3], cr[3])
```

Table 8-1 shows quantum circuits for the unitary transformations A1-3, B1-3 for IBM Q Experience composer.

Table 8-1. *Quantum circuits for the unitary transformations in Listing 8-6*

Transformation	Circuit
$A1 = \dfrac{1}{\sqrt{2}} \begin{bmatrix} i & 0 & 0 & 1 \\ 0 & -i & 1 & 0 \\ 0 & i & 1 & 0 \\ 1 & 0 & 0 & i \end{bmatrix}$	
$A2 = \dfrac{1}{2} \begin{bmatrix} i & 1 & 1 & i \\ -i & 1 & -1 & i \\ i & 1 & 1 & -i \\ -i & 1 & -1 & -i \end{bmatrix}$	
$B1 = \dfrac{1}{2} \begin{bmatrix} i & -i & 1 & 1 \\ -i & -i & 1 & -1 \\ 1 & 1 & -i & i \\ -i & i & 1 & 1 \end{bmatrix}$	
$B2 = \dfrac{1}{2} \begin{bmatrix} -1 & i & 1 & i \\ 1 & i & 1 & -i \\ 1 & -i & 1 & i \\ -1 & -i & 1 & -i \end{bmatrix}$	
$B3 = \dfrac{1}{\sqrt{2}} \begin{bmatrix} 1 & 0 & 0 & 1 \\ -1 & 0 & 0 & 1 \\ 0 & 1 & 1 & 0 \\ 0 & 1 & -1 & 0 \end{bmatrix}$	

In Table 8-1 note that A3 is not included because the composer does not support the swap gate required by Listing 8-5. This does not mean the quantum program can't be run in the simulator or real device, however. It simply means the circuit cannot be created in the composer. Thus, for the final step, Alice and Bob measure their qubits in the computational basis.

Measure in the Computational Basis

After measurement, Alice and Bob end up with two bits each which represent their respective outputs. To obtain the third bit, and thus a final answer, they apply their parity rules. That is, Alice's sum must be even, and Bob's must be odd. For example, for a = 2, b = 3 (see Table 8-2):

$$(A2 \otimes B3)|\Psi\rangle = \frac{1}{2\sqrt{2}}\Big[|0000\rangle - |0010\rangle - |0101\rangle + |0111\rangle + |1001\rangle + |1011\rangle - |1100\rangle - |1110\rangle\Big]$$

Table 8-2. *Answer permutations for a = 2, b =3 of the magic square*

Ψ	Alice's answer	Bob's answer	Square
|0000>	000	001	$\begin{bmatrix} & & 0 \\ 0 & 0 & 0 \\ & & 1 \end{bmatrix}$
|0010>	000	100	$\begin{bmatrix} & & 1 \\ 0 & 0 & 0 \\ & & 0 \end{bmatrix}$
|0101>	011	010	$\begin{bmatrix} & & 0 \\ 0 & 1 & 1 \\ & & 0 \end{bmatrix}$

(continued)

Table 8-2. (*continued*)

Ψ	Alice's answer	Bob's answer	Square
|0111>	011	111	$\begin{bmatrix} & & 1 \\ 0 & 1 & 1 \\ & & 1 \end{bmatrix}$
|1001>	101	010	$\begin{bmatrix} & & 0 \\ 1 & 0 & 1 \\ & & 0 \end{bmatrix}$
|1011>	101	111	$\begin{bmatrix} & & 1 \\ 1 & 0 & 1 \\ & & 1 \end{bmatrix}$
|1100>	110	001	$\begin{bmatrix} & & 0 \\ 1 & 1 & 0 \\ & & 1 \end{bmatrix}$
|1110>	110	101	$\begin{bmatrix} & & 1 \\ 1 & 1 & 0 \\ & & 1 \end{bmatrix}$

Listing 8-7 shows a section of the script to loop through all the rounds of the magic square:

- It loops through *a [1,3]* and *b[1,3]* inclusive.

- For each (a, b), a circuit for Alice (Alice-a) and a circuit for Bob (Bob-b) are retrieved from Listing 8-5.

- The shared entangled state Ψ, Alice-a and Bob-b circuits are submitted for execution to the simulator or real quantum device.

- Two bits are extracted for Alice and two for Bob from the answer such as {'0011': 1}.

- The parity rules are applied: Alice's sum must be even, and Bob's sum must be odd.

- Finally, the answer is verified, and the winning probability is displayed.

Listing 8-7. Script for all rounds of the magic square

```
def all_rounds(backend, real_dev, shots=10):
  nWins = 0
  nLost = 0
  for a in range(1,4):
    for b in range(1,4):
      print("For a = " + str(a) + " , b = " + str(b))
      rWins = 0
      rLost = 0

      aliceCircuit  = aliceCircuits["Alice" + str(a)]
      bobCircuit    = bobCircuits["Bob" + str(b)]
      circuitName   = "Alice" + str(a) + "Bob"+str(b)

      root = sharedEntangled.compose(aliceCircuit).compose(bobCircuit)

      if real_dev:
        device_cfg      = backend.configuration()
        device_coupling = device_cfg.coupling_map
        results = execute([root], backend=backend, shots=shots
              , coupling_map=device_coupling).result()
      else:
        results = execute([root], backend=backend, shots=shots).result()

      answer = results.get_counts()

      for key in answer.keys():
        kfreq = answer[key] #frequencies of keys obtained from measurements
        aliceAnswer = [int(key[-1]), int(key[-2])]
        bobAnswer   = [int(key[-3]), int(key[-4])]
```

```python
        if sum(aliceAnswer) % 2 == 0:
          aliceAnswer.append(0)
        else:
          aliceAnswer.append(1)
        if sum(bobAnswer) % 2 == 1:
          bobAnswer.append(0)
        else:
          bobAnswer.append(1)

        print(str(backend) + " answer: " + key + " Alice: " +
        str(aliceAnswer)
          + " Bob:" + str(bobAnswer))

        if(aliceAnswer[b-1] != bobAnswer[a-1]):
          #print(a, b, "Alice and Bob lost")
          nLost += kfreq
          rLost += kfreq
        else:
          #print(a, b, "Alice and Bob won")
          nWins += kfreq
          rWins += kfreq
      print("\t#wins = ", rWins, "out of ", shots, "shots")

  print("Number of Games = ", nWins+nLost)
  print("Number of Wins = ", nWins)
  print("Winning probabilities = ", (nWins*100.0)/(nWins+nLost))

################################################
# main
################################################
if __name__ == '__main__':

  name = "qasm_simulator"
  backend = Aer.get_backend(name)

  real_dev = False
  all_rounds(backend, real_dev)
```

A run of the full script at Workspace\Ch08\p_magicsq.py against the remote simulator is shown in listing 8-8.

Listing 8-8. Simplified standard output from a run of all rounds of the magic square

```
c:\python36-64\python.exe p_magicsq.py
For a = 1 , b = 1
ibmq_qasm_simulator answer: 1000 Alice: [0, 0, 0] Bob:[0, 1, 0]
ibmq_qasm_simulator answer: 1010 Alice: [0, 1, 1] Bob:[0, 1, 0]
ibmq_qasm_simulator answer: 1111 Alice: [1, 1, 0] Bob:[1, 1, 1]
ibmq_qasm_simulator answer: 0111 Alice: [1, 1, 0] Bob:[1, 0, 0]
ibmq_qasm_simulator answer: 0000 Alice: [0, 0, 0] Bob:[0, 0, 1]
ibmq_qasm_simulator answer: 0101 Alice: [1, 0, 1] Bob:[1, 0, 0]
        #wins =  10 out of  10 shots
For a = 1 , b = 2
ibmq_qasm_simulator answer: 1000 Alice: [0, 0, 0] Bob:[0, 1, 0]
ibmq_qasm_simulator answer: 1001 Alice: [1, 0, 1] Bob:[0, 1, 0]
ibmq_qasm_simulator answer: 1111 Alice: [1, 1, 0] Bob:[1, 1, 1]
ibmq_qasm_simulator answer: 0110 Alice: [0, 1, 1] Bob:[1, 0, 0]
ibmq_qasm_simulator answer: 0000 Alice: [0, 0, 0] Bob:[0, 0, 1]
ibmq_qasm_simulator answer: 0001 Alice: [1, 0, 1] Bob:[0, 0, 1]
        #wins =  10 out of  10 shots
...
For a = 3 , b = 3
ibmq_qasm_simulator answer: 1000 Alice: [0, 0, 0] Bob:[0, 1, 0]
ibmq_qasm_simulator answer: 1011 Alice: [1, 1, 0] Bob:[0, 1, 0]
ibmq_qasm_simulator answer: 1101 Alice: [1, 0, 1] Bob:[1, 1, 1]
ibmq_qasm_simulator answer: 1110 Alice: [0, 1, 1] Bob:[1, 1, 1]
ibmq_qasm_simulator answer: 0111 Alice: [1, 1, 0] Bob:[1, 0, 0]
ibmq_qasm_simulator answer: 0010 Alice: [0, 1, 1] Bob:[0, 0, 1]
        #wins =  10 out of  10 shots
Number of Games =  90
Number of Wins =  90
Winning probability =  100.0
```

Note   If running in a real device, the winning probability will not be 100% due to environmental noise and gate error.

Answers for the Mermin-Peres Magic Square Exercise

1. A magic square whose row product is even and whose column product is odd is given here. Note that such a square is not possible due to the odd number of cells.

-1	-1	1
-1	1	-1
-1	1	?

2. The permutation table for the square in answer 1 is:

N	a	b	Alice	Bob	Intersection	Win/Loose
1	1	1	-1,-1,1	-1,-1,-1	-1/-1	W
2	1	2	-1,-1,1	-1,1,1	-1/-1	W
3	1	3	-1,-1,1	1,-1,? (1)	1/1	W
4	2	1	-1,1,-1	-1,-1,-1	-1/-1	W
5	2	2	-1,1,-1	-1,1,1	1/1	W
6	2	3	-1,1,-1	1,-1,? (1)	-1/-1	W
7	3	1	-1,1,? (-1)	-1,-1,-1	-1/-1	W
8	3	2	-1,1,? (-1)	-1,1,1	1/1	W
9	3	3	-1,1,? (-1)	1,-1,? (1)	-1/1	L

3. Note that, in the previous step for rows 7-9, Alice's answer must
    be -1 so the product can be even (1). Plus, in columns 3, 6, and 9,
    Bob's answer must be 1 so his product can be odd (-1). Finally, the
    probability is calculated by dividing the total number of wins by
    the total number of permutations. Thus:

$$P = \frac{\sum W}{N} = \frac{8}{9} = 88.88\%$$

In this chapter, you have learned how the power of quantum entanglement can provide significant speedups over classical computation. With a quantum beam balance, it is possible to achieve quartic speedups for classical puzzles like the counterfeit coin problem. For others, such as the magic square, entanglement gives quasi-magical telepathy among players. All in all, this chapter has shown how quantum mechanics is as confusing, bizarre, and fascinating as always. It never fails to deliver.

In the next chapter, you will learn about arguably the most famous quantum algorithm of them all: the notorious Shor's integer factorization. An algorithm that may smash asymmetric cryptography!

Quantum Advantage with Deutsch-Jozsa, Bernstein-Vazirani, and Simon's Algorithms

In this chapter, we study three algorithms of little practical use but important because they were the first to show that quantum computers can solve problems significantly faster than classical ones. Consider the time complexities O(n) for the algorithms: classical vs. quantum in Table 9-1 (where n is the size of the input).

Table 9-1. *Time complexities for Deutsch-Jozsa, Bernstein-Vazirani, and Simon algorithms*

Name	Classical	Quantum
Deutsch-Jozsa	$2^{n-1} + 1$	1
Bernstein-Vazirani	n	1
Simon	$2^{n/2}$	n

© Vladimir Silva 2024
V. Silva, *Quantum Computing by Practice*, https://doi.org/10.1007/978-1-4842-9991-3_9

Besides the significant performance boost, these algorithms illustrate fundamental concepts that apply to virtually all algorithms out there:

- Massive parallelism: By applying the Hadamard gate to all inputs, we create a set of all binary permutations, and most incredible of all, they are all consumed at the same time. This may not sound like much, but consider a 40 qubit processor: It can consume $2^{40} = 1TB$ of binary permutations at once. That is an astounding amount of raw power that no supercomputer in existence can match.

- Oracles: These are black boxes that perform some transformation on the binary permutations. When you have a huge permutation space, the oracle gives the means of changing most of them so they can cancel each other. Virtually all quantum algorithms will use some kind of oracle.

- Phase kickback: This is an extremely useful concept that you must fully understand to build quantum algorithms. Phase kickback essentially changes the phase (sign) of some permutations so they can cancel each other (after being transformed by the oracle). The result of this process is a reduced permutation space that will contain the solution to the problem. But more about this in the next section.

In these three concepts lies the key to quantum advantage: imagine you have a binary permutation space of 2^{40} created using Hadamard gates of your qubits. Within this space lies a single permutation waiting to be found giving the solution to the problem. We need a mechanism for altering some permutations so they can cancel each other (the oracle). Also, we need a way to change their phase (phase kickback) so that when we bring them out of superposition they will cancel by addition (remember that superposition is simply the addition of quantum states). The final measurement will trigger this process resulting in a reduced permutation set that gives the solution to the problem; the hardest part of all this is to encode the problem in such a binary permutation space, and that is where all research takes place. So let's take a look at this quintessential part of quantum information science: we start with the extremely important concept of phase kickback.

Phase Kickback

Phase kickback occurs when the eigenvalue added by a gate to a qubit is "kicked back" into a different qubit via a controlled operation. For this to occur, the qubits must be in superposition: Remember that:

$$H|0\rangle = |+\rangle, \quad H|1\rangle = |-\rangle, \quad X|-\rangle = -|-\rangle$$

In a 2 qubit state, when the control qubit is not in superposition |0> or |1>, the phase is global and has no observable effect (global phases are irrelevant), Thus, applying the Control-X operator gives:

$$CX|-0\rangle = X|-\rangle \otimes I|0\rangle = |-0\rangle, \quad CX|-1\rangle = X|-\rangle \otimes I|1\rangle = -|-1\rangle$$

Note that we are using **Qiskit little endian bit ordering** (the control or least significant goes on the right). Phase kick back occurs when the control (right) qubit is in superposition. The control qubit in state |1> applies a phase factor to the corresponding target qubit. This applied phase factor in turn introduces a relative phase into the control qubit thus:

$$CX|-+\rangle = \frac{1}{\sqrt{2}}\left(CX|-0\rangle + CX|-1\rangle\right) - \frac{1}{\sqrt{2}}\left(|\ 0\rangle \mid 1\rangle\right)$$

$$CX|-+\rangle = |-\rangle \otimes \frac{1}{\sqrt{2}}\left(|0\rangle - |1\rangle\right) = |--\rangle$$

Let's put this to the test with a simple lab exercise.

Exercise 9.1: Start the Quantum Lab to quickly create a circuit for the 2 qubit state |-+>, and display the Bloch spheres to visualize the rotations using the state vector simulator. Note that qubit 0 will be in state (+).

```
from qiskit import *
from qiskit.visualization import *

qc = QuantumCircuit(2)
qc.h(0)
qc.x(1)
qc.h(1)

display (qc.draw())
backend = Aer.get_backend("statevector_
simulator")
result = execute(qc, backend).result()

plot_bloch_multivector(result.get_
statevector())
```

Exercise 9.2: Add a Control-X (0,1) operator to the circuit to visualize how the negative phase is kicked back from qubit-1 (target) to qubit-0 (control in superposition) resulting in the state |-->

```
from qiskit import *
from qiskit.visualization import *

qc = QuantumCircuit(2)
qc.h(0)
qc.x(1)
qc.h(1)
qc.cx(0,1)

display (qc.draw())
backend = Aer.get_backend("statevector_
simulator")
result = execute(qc, backend).result()

plot_bloch_multivector(result.get_
statevector())
```

Kickback with Arbitrary Phases

In quantum mechanics, phase and sign flips are used interchangeably. For example, a negative phase kickback is a rotation of pi degrees over the Bloch sphere. This is a powerful concept that allows for phase kickbacks of arbitrary angles between qubits provided that we use a controlled operation and the target qubit is in superposition. Such a feat can be achieved using the handy phase gate which performs a rotation over the Z axis of λ (lambda) degrees: $P(\lambda) = \begin{bmatrix} 1 & 0 \\ 0 & e^{i\lambda} \end{bmatrix}$.

Let's see how this is done. Consider Listing 9-1. We start with two qubits in the state |+1> (left side of Figure 9-1). We then apply a controlled phase of pi/4 degrees (T-gate). This has the effect of a phase kickback of a quarter turn (over the Z axis) from the control qubit-1 to the target qubit-0 which is in superposition (right side of Figure 9-1).

Listing 9-1. Phase kickback using the T-gate

```
from qiskit import *
from qiskit.visualization import *
from math import pi

qc = QuantumCircuit(2)

qc.h(0)
qc.x(1)

# Add Controlled-T gate:
# 1) Comment this to view the left side of figure 9-1.
# 2) Uncomment to view the right side of figure 9-1.
#qc.cp(pi/4, 0, 1)

display (qc.draw())
backend = Aer.get_backend("statevector_simulator")
result = execute(qc, backend).result()

plot_bloch_multivector(result.get_statevector())
```

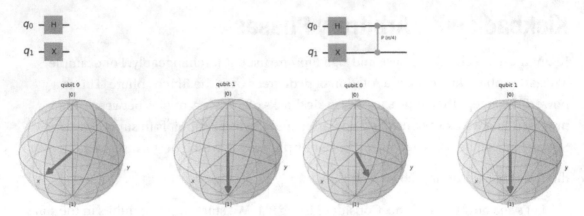

Figure 9-1. *Phase kickback using the T-gate*

Exercise 9.3: What would be the position of qubit-0 if we apply a rotation of –pi/2 (S-dagger)? Tip: Given that S rotates by pi/2 degrees over the Z-axis, update the rotation angle of the qc.cp() instruction.

Phase kickback is used in all three algorithms studied in the next sections.

Deutsch-Jozsa

This algorithm was introduced in 1992 by David Deutsch and Richard Jozsa.[1] Given a hidden Boolean function f({x0,x1,..,xn}) where {x0,x1,..,xn} is a string of bits, it returns 0 if the function is constant or 1 if balanced. A Boolean constant function returns all 0's or 1's for any input. A Boolean balanced function returns 0's for half the inputs and 1's for the other half. The task is to determine whether the given function is balanced or constant.

Tip   Examples of balanced Boolean functions are the function that copies the first bit of its input to the output, and the XOR (exclusive or – verify this by looking at its truth table). Balanced Boolean functions are primarily used in cryptography.

---

[1] David Deutsch and Richard Jozsa (1992). "Rapid solutions of problems by quantum computation". Proceedings of the Royal Society of London A. 439: 553–558. doi:10.1098/rspa.1992.0167

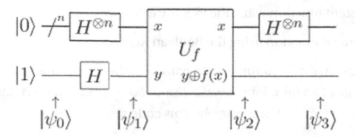

Figure 9-2. *Deutsch-Jozsa quantum circuit*

The circuit in Figure 9-2 can solve the problem with 100% confidence after a single call. Let's see how this is done. We start with:

$$|\Psi_0\rangle = |0\rangle^{\otimes n}|1\rangle$$

The Hadamard gates put the input bits in superposition and the ancillary qubit|1> in |->:

$$|\Psi_1\rangle = \frac{1}{\sqrt{2^{n+1}}}\sum_{x=0}^{2^n-1}|x\rangle|-\rangle$$

where x is the input bit string and next the function U_f is the oracle which uses phase kickback to transform the state $|x\rangle|y\rangle \rightarrow |x\rangle|y \oplus fx\rangle$

$$|\Psi_2\rangle = \frac{1}{\sqrt{2^{n+1}}}\sum_{x=0}^{2^n-1}|x\rangle(|fx\rangle - |1 \oplus fx\rangle) = \frac{1}{\sqrt{2^{n+1}}}\sum_{x=0}^{2^n-1}(-1)^{fx}|x\rangle(|0\rangle - |1\rangle)$$

Next, the ancillary qubit is discarded; apply the Hadamard to the first register; we obtain:

$$|\Psi_2\rangle = \frac{1}{2^n}\sum_{y=0}^{2^n-1}\left[\sum_{x=0}^{2^n-1}(-1)^{fx}(-1)^{x.y}\right]|y\rangle$$

where x.y is the bitwise product of the input bit string x and the ancillary qubit (y).

Finally, after measuring the first register, the probability becomes $\Pr = \left|\sum_{x=0}^{2^n-1}(-1)^{fx}\right|^2$ which evaluates to 1 if f(x) is balanced. Let's implement this algorithm for a 3 qubit circuit with a 2-bit size -x and 1-bit size-y for two functions:

1. A constant function using the identity gate

2. A balanced function using the Boolean XOR

Run Listing 9-2 in the Quantum Lab to construct the Deutsch-Jozsa circuit for a 2 bit function; it contains two functions, *constant2* and *balanced2*, which create the oracles for a constant and balanced f(x), respectively. This code starts by

1. Putting the input qubits (x) in superposition.

2. It then applies the oracle for the constant or balanced function.

3. It moves the inputs back to the computational basis.

4. Measures the input x. If 0 the function is constant, else it is balanced.

Listing 9-2. Deutsch-Jozsa for 2 qubits with constant or balanced functions

```
from qiskit import QuantumRegister, ClassicalRegister, QuantumCircuit
from qiskit import Aer, execute
from qiskit.tools.visualization import plot_histogram

# Create quantum/classical registers and a quantum circuit
qin     = QuantumRegister(2, 'x')
qout    = QuantumRegister(1, 'y')
c       = ClassicalRegister(2)
qc      = QuantumCircuit(qin,qout,c)

# This is  a balanced function XOR
def balanced2(circ, input, output) :
    circ.cx(input[0],output)
    circ.cx(input[1],output)

# This is a constant function that doesn't change anything
def constant2(circ, input, output) :
    circ.i(output)

# Build the Deutsh-Josza circuit
# First apply phase shift to negate where f(x) == 1
qc.h(qin)
```

```
qc.x(qout)
qc.h(qout)

# balanced2() for balanced or constant2() for constant
constant2(qc,qin,qout)
#balanced2(qc,qin,qout)

qc.h(qout)
qc.barrier()  # barrier to make circuit look nicer

# Then Walsh-Hadamard on input bits
qc.h(qin)

# For constant function, output is |00> with probability 1
# For balanced function, output is |11> with probability 1
qc.measure(qin,c)

# display only works in Quantum Lab
display(qc.draw())

# Simulate and show results
backend     = Aer.get_backend('qasm_simulator')
job = execute(qc, backend, shots-512)
result  = job.result()

plot_histogram(result.get_counts())
```

Run Listing 9-2 in the Quantum Lab to obtain the result shown in Figure 9-3.

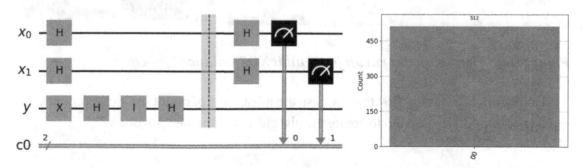

Figure 9-3. *Deutsch-Jozsa circuit and result for a constant function*

Exercise 9.4: Modify Listing 9-2 to use the balanced XOR function instead. Run in the Quantum Lab and verify the circuit and results histogram matches Figure 9-4.

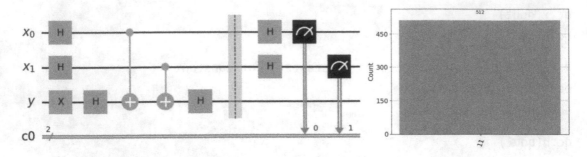

Figure 9-4. *Deutsch-Jozsa circuit and result for the XOR balanced function*

Exercise 9.5: Modify Listing 9-2 to generalize the Deutsch-Jozsa algorithm for any number of qubits. Run with a balanced XOR function for 3 qubits and verify the result matches Figure 9-5. Tip: Set a variable n = 3, then simply update qin = QuantumRegister(n, 'x'). Finally, update the balanced XOR function with a *for* loop from 0 to n:

```
def balanced(n, circ, input, output) :
  for i in range (n):
    circ.cx(input[i],output)
```

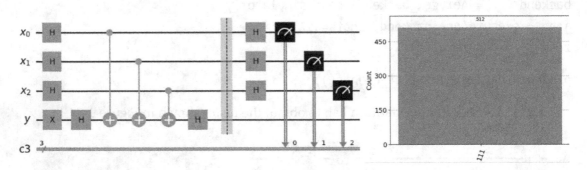

Figure 9-5. *Deutsch-Jozsa circuit for 3 qubit XOR balanced function*

Deutsch-Jozsa was the first to showcase the advantage of quantum. Researchers quickly built upon this design to create the algorithms in the next sections.

298

Bernstein-Vazirani (BV)

Bernstein-Vazirani is an extension of the Deutsch-Jozsa algorithm introduced in 1997.[2] The algorithm is similar yet more complex than Deutsch-Jozsa:

1. We start with the same function $f(\{x0,x1,..,xn\})$ where $\{x0,x1,..,xn\}$ is a string of bits, $f(x)$ returns 0 or 1.

2. Here is the difference: Instead of the function being balanced or constant as in Deutsch-Jozsa, $f(x)$ returns the bitwise product of the input x with some secret string s.

3. The goal is to find the secret string s for which $f(x) = s \cdot x \pmod 2$ using the transformation $H^{\otimes n}|s\rangle = \frac{1}{\sqrt{2^n}} \sum_{x \in \{0,1\}} (-1)^{s.x}|x\rangle$

where |s> is an n–qubit state permutation of the secret bit string and |x> is the input space. The circuit and mathematical explanation are shown in Figure 9-6.

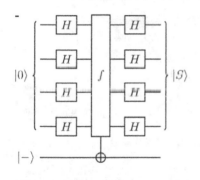

Figure 9-6. *Quantum circuit for Bernstein-Vazirani*

The Hadamard applied at two qubits gives:

$$H^{\otimes 2}|00\rangle = \frac{1}{2}\left(|00\rangle + |01\rangle + |01\rangle + |11\rangle\right)$$

$$H^{\otimes 2}|01\rangle = \frac{1}{2}\left(|00\rangle - |01\rangle + |01\rangle - |11\rangle\right)$$

$$H^{\otimes 2}|10\rangle = \frac{1}{2}\left(|00\rangle + |01\rangle - |01\rangle - |11\rangle\right)$$

$$H^{\otimes 2}|11\rangle = \frac{1}{2}\left(|00\rangle - |01\rangle - |01\rangle + |11\rangle\right)$$

This can be expressed as a sum over the input string of bits x:

$$H^{\otimes 2}|a\rangle = \frac{1}{2} \sum_{x \in \{0,1\}} (-1)^{a.x}|x\rangle$$

[2] Ethan Bernstein and Umesh Vazirani (1997) "Quantum Complexity Theory" SIAM Journal on Computing, Vol. 26, No. 5: 1411-1473, doi:10.1137/S0097539796300921

Tip   Remember that $H|0\rangle = \frac{1}{\sqrt{2}}(|0\rangle + |1\rangle)$, $H|1\rangle = \frac{1}{\sqrt{2}}(|0\rangle - |1\rangle)$, $|0\rangle \otimes |0\rangle = |00\rangle$.

Exercise 9.6: Show by multiplication that $H^{\otimes 2}|00\rangle = \frac{1}{2}(|00\rangle + |01\rangle + |01\rangle + |11\rangle)$.

$$H^{\otimes 2}|00\rangle = \frac{1}{\sqrt{2}}(|0\rangle + |1\rangle) \otimes \frac{1}{\sqrt{2}}(|0\rangle + |1\rangle) = \frac{1}{2}(|00\rangle + |01\rangle + |01\rangle + |11\rangle)$$

Exercise 9.7: Show by multiplication that $H^{\otimes 2}|01\rangle = \frac{1}{2}(|00\rangle - |01\rangle + |01\rangle - |11\rangle)$

A generalized algorithm for any secret string size is shown in Listing 9-3, the circuit and results in Figure 9-7.

Listing 9-3. Bernstein-Vazirani algorithm generalized for any hidden string (s)

```
# initialization
import matplotlib.pyplot as plt
import numpy as np
from qiskit import *
from qiskit.visualization import *

# generator for a hidden function with n qubits as input
def bv_oracle (n, s) : #, qc) :
    # reverse s to fit qiskit's qubit ordering
    s   = s[::-1]
    qc  = QuantumCircuit(n + 1)
    for q in range(n):
        if s[q] == '0':
            qc.i(q)
        else:
            qc.cx(q, n)
    return qc

# n-qubit input version for the Bernstein-Vazirani (BV) Algorithm
def bv_circuit (n, s):
    # We need a circuit with n qubits, plus one auxiliary qubit
    # Also need n classical bits to write the output to
```

```
    bv = QuantumCircuit( n + 1, n)

    # put auxiliary in state |->
    #bv.x(n)
    bv.h(n)
    bv.z(n)

    # Apply Hadamard gates before querying the oracle
    for i in range(n):
        bv.h(i)

    # Apply barrier
    bv.barrier()

    # Apply the inner-product oracle
    bv = bv.compose(bv_oracle (n, s))

    # Apply barrier
    bv.barrier()

    #Apply Hadamard gates after querying the oracle
    for i in range(n):
        bv.h(i)

    # Measurement
    for i in range(n):
        bv.measure(i, i)

    return bv

# Main
s           = '101'      # the hidden binary string
n           = len(s)      # number of qubits used to represent s

bvc         = bv_circuit (n, s)

# Run in simulator
backend     = Aer.get_backend('qasm_simulator')
results      = execute(bvc, backend).result()
answer      = results.get_counts()
```

```
display (bvc.draw())
print ('s =',s , ' n =',n , ' Result =', answer)

plot_histogram(answer)
```

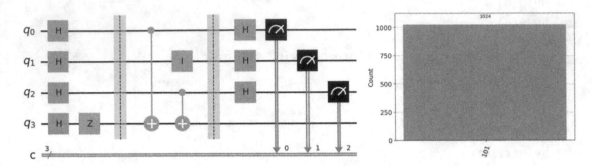

Figure 9-7. *Bernstein-Vazirani circuit and results for hidden bit string 101*

Exercise 9.8: Run listing 9-3 in the Quantum Lab for different secret strings of multiple sizes and permutations. Verify that the results are correct.

Exercise 9.9: How would the oracle circuits look for secret strings 011 and 110? Tip: Use the Quantum Lab and remember that Qiskit uses little-endian bit ordering (the least significant bit draws at the top of the circuit).

Simon's Algorithm

Simon's algorithm was introduced in 1997[3] and was the first quantum algorithm to show an exponential speed-up versus the best classical solution. As a matter of fact, this work inspired the notorious Quantum Fourier Transform, which is the groundwork for the famous Shor's factorization algorithm. Just like the previous algorithms, we are given an oracle f(x) which maps an input to its output in two ways:

- One to one: It maps exactly one output for every input. For example, f(0) = 0, f(1) = 1, f(2) = 2, f(3) = 3

- Two to one: It maps two inputs to every output. For example, f(0) = 0, f(1) = 1, f(2) = 0, f(3) = 1

[3] Daniel R. Simon (1997) "On the Power of Quantum Computation" SIAM Journal on Computing, 26(5), 1474–1483, doi:10.1137/S0097539796298637

The catch is that there is a hidden bit string b where $f(x1) = f(x2)$ given $x1 \oplus x2 = b$. The task is to find the hidden bit string b such that $b = x1x2x3...$ represents a one-to-one mapping of $f(x)$. Let's take a look at the internals in Figure 9-8.

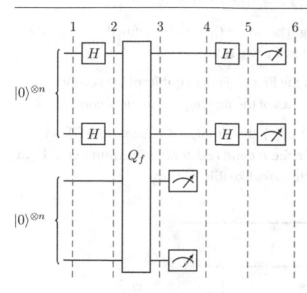

Figure 9-8. *Quantum circuit for Simon's algorithm*

1) Two n-qubit registers are initialized.
$$|\psi 1\rangle = |0\rangle^{\otimes n}|0\rangle^{\otimes n}$$

2) Apply a Hadamard to the 1st register:

$$|\Psi 2\rangle = \frac{1}{\sqrt{2^n}} \sum_{x \in \{0,1\}} |x\rangle|0\rangle^{\otimes n}$$

3) Apply the oracle Qf:

$$|\Psi 3\rangle = \frac{1}{\sqrt{2^n}} \sum_{x \in \{0,1\}} |x\rangle|f(x)\rangle$$

4) Measure the 2nd register:

$$|\Psi 4\rangle = \frac{1}{\sqrt{2}} (|x\rangle + |y\rangle), y = x \oplus b$$

5) Apply the Hadamard to the 1st register.

$$|\Psi 5\rangle = \frac{1}{\sqrt{2^{n+1}}} \sum_{x \in \{0,1\}} \left[(-1)^{x.z} + (-1)^{y.z} \right]|z\rangle$$

6) Measure the 1st register:
$$(-1)^{x.z} = (-1)^{y.z}$$

Finally, by executing this circuit multiple times and solving $x.z = y.z$ where $y = x \oplus b$, we end up with a system of equations from which the secret string b can be found:

$$\begin{cases} b.z1 = 0 \\ b.z2 = 0 \\ \quad ... \\ b.zn = 0 \end{cases}$$

Let's take a look at how this works for b = 10:

1. The two registers are initialized to the zero state $|\psi 1\rangle = |00\rangle|00\rangle$.

2. Apply the Hadamard to the first register. This builds all permutations for 2 qubits. Thus, $|\Psi 2\rangle = \frac{1}{2}(|00\rangle + |01\rangle + |10\rangle + |11\rangle)$.

3. Here is the key to the whole thing. The oracle Q_f must be defined for b = 10. To do so follow the following two rules:

    a. Apply CX gates from qubits of the first register to qubits of the second register. We do it to copy the states of the first register to the second register.

    b. Find the index (i) for the first 1 in (s), then apply n-CX gates from (i) to the indices of each 1 occurrence in the second register (n is the number of 1's in s). Thus for string 10, the oracle looks like this:

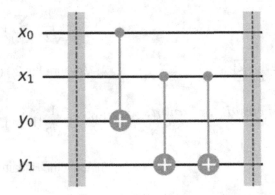

$$|\Psi 3\rangle = \frac{1}{2}\left(\begin{array}{l} |00\rangle|0\oplus 0\oplus 0,|0\oplus 0\oplus 0\rangle\rangle + |01\rangle|0\oplus 0\oplus 1,|0\oplus 0\oplus 1\rangle\rangle \\ +|10\rangle|0\oplus 1\oplus 0,|0\oplus 1\oplus 0\rangle\rangle + |11\rangle|0\oplus 1\oplus 1,|0\oplus 1\oplus 1\rangle\rangle \end{array} \right)$$

$$|\Psi 3\rangle = \frac{1}{2}(|00\rangle|00\rangle + |01\rangle|11\rangle + |10\rangle|11\rangle + |11\rangle|00\rangle)$$

4. If we measure the 2nd register, we get 50% probabilities for either

    00 or 11. If, for example, we measure 11, then $|\Psi 4\rangle = \frac{1}{\sqrt{2}}(|01\rangle + |10\rangle)$.

5. After Hadamard in the 1st register:

$$|\Psi 5\rangle = \frac{1}{2}(|00\rangle - |01\rangle - |10\rangle + |11\rangle + |00\rangle + |01\rangle - |11\rangle - |11\rangle) = (|00\rangle - |10\rangle)$$

6. Finally, measure the 1st register to get $|00\rangle - |10\rangle$ with equal probability. This gives the system of equations $\begin{cases} b.00 = 0 \\ b.10 = 0 \end{cases}$.

7. Solve the system of equations to obtain $b2(1) + b1(0) = 0$, $b2 = 1$, $b1 = 0$. Thus $b = b2b1 = 10$.

Listing 9-4 shows the generalized algorithm for an n-qubit string (b). Its output is shown in Figure 9-9.

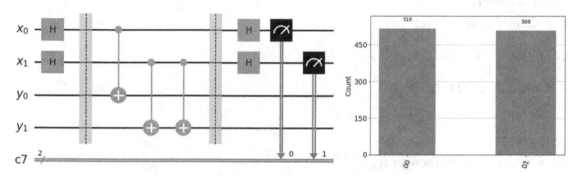

Figure 9-9. *Simon's algorithm circuit with output for b = 10*

Note   Listing 9-4 uses the qiskit textbook package which is not a standard module; it has to be installed separately with the command: pip install git+https://github.com/qiskit-community/qiskit-textbook. git#subdirectory=qiskit-textbook-src.

Listing 9-4. Simon's algorithm for an n-qubit arbitrary string (b)

```
from qiskit import *
from qiskit.visualization import *
from qiskit_textbook.tools import simon_oracle

def dotprod(a, b):
    temp_sum = 0
    for i in range(len(a)):
        temp_sum += int(a[i])*int(b[i])
    return temp_sum % 2
```

```python
# Main function
b = '10'
n = len(b)

# 1st register
x = QuantumRegister(n, 'x')
# 2nd register
y = QuantumRegister(n, 'y')
# measurement
c = ClassicalRegister (n)

qc = QuantumCircuit(x,y, c)

# H in 1st register
qc.h(x)

qc.barrier()
qc = qc.compose(simon_oracle(b))
qc.barrier()

# H in 1st register
qc.h(x)

# Measure  1st register
qc.measure(x, c)
display(qc.draw())

backend = Aer.get_backend('qasm_simulator')
job = execute(qc, backend=backend)
counts = job.result().get_counts()

display(plot_histogram(counts))

for code in counts:
    print("(" + str(b) + " dot " + str(code) + ")%2 = "
    + str(dotprod(b, code)))
```

Exercise 9.10: Run Listing 9-4 in the Quantum Lab and construct Simon oracles for the three bit strings 110, 101.

Rules for Simon Oracle Construction

Try to keep these two rules handy when building a Simon oracle. They are the key to the whole algorithm:

1. Copy the states of the first register to the second register. That is, apply CX gates from all qubits of the first register to the corresponding qubit of the second register.

2. Find the index (i) of the first 1 in (s), then apply **n-CX** gates from (i) to the indices of each occurrence in the second register where n is the number of 1's in the string. This is equivalent to $|x\rangle \rightarrow |s.x\rangle$ *if* $qubit(i) = = 1$.

From these rules, we can see that the total number of CX gates equals the number of qubits in the string plus the number of 1s. Let's take a closer look. Also, note that rule one applies to all permutations.

Dissecting Simon's Oracle

In this section we'll practice building Simon oracles for 3-bit strings: Table 9-2 shows some of the permutations, along with the CX gates required by the oracle. Remember that, Qiskit uses little-endian bit ordering; the zero-index starts at the right of the string (bottom of the circuit). Also, remember that all permutations start with the sequence CX(0,0) CX(1,1) CX(2,2).

Table 9-2. *Simon oracles for a 3-bit string*

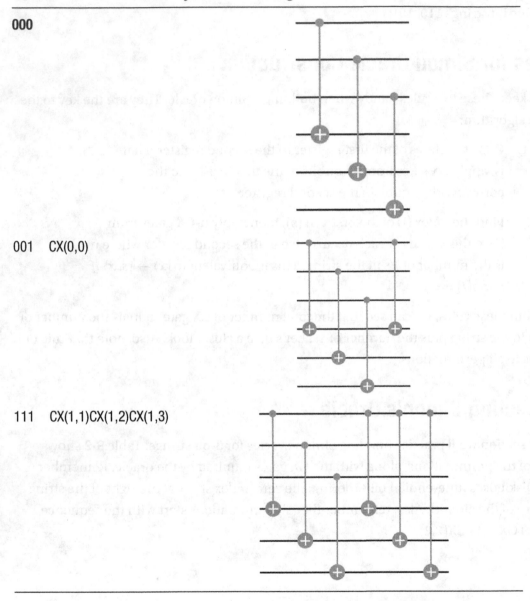

000

001   CX(0,0)

111   CX(1,1)CX(1,2)CX(1,3)

Exercise 9.11: Complete the gate sequence for the 3-bit permutations of Simon's oracle.

000      $CX(0,0)CX(1,1)CX(2,2)$ 100   ?

001    $CX(0,0)CX(1,1)CX(2,2)CX(0,0)$101   ?

010    $CX(0,0)CX(1,1)CX(2,2)CX(1,1)$ 110   ?

011   ?   111 $CX(0,0)CX(1,1)CX(2,2)CX(0.0)CX(0,1)CX(0,2)$

Exercise 9.12: Write a python function to build a Simon oracle for any bit string. Test your function in the Quantum Lab to visualize oracles and verify they are correct. Tip: Use the following skeleton which already implements rule 1 (add rule 2). You can compare your function with the official Qiskit's *simon_oracle(b)* from package *qiskit_textbook.tools*.

```
# n-qubit version for Simon's oracle
def oracle (s):
  # reverse b for qiskit's qubit ordering
  s   = s[::-1]
  n   = len(s)
  qc  = QuantumCircuit(n * 2)

  # all 0s, so just exit
  if '1' not in s:
    return qc

  # index of first non-zero bit in s
  i = s.find('1')

  for q in range(n):
    # Rule1: Copy; |x>|0> -> |x>|x>
    qc.cx(q, q+n)

    # Add Rule 2: |x> -> |s.x> if q(i) == 1
    # ...
  return qc
```

Extended Practice Exercises

9.13 Select the involutory gates in the following list (Select 3): S, T, X, Y, H
(Tip: Involutory gates equal their own inverse $M = M^{-1}$).

9.14 T-gate is a phase gate with what value of the phase?

9.15 Which of the following code snippets for the following quantum circuit will put the given qubits in an equiprobable state (equal probability of being in all states).

```
qc=QuantumCircuit(2)
a) qc.h(0) qc.cx(0,1) b) qc.h(0) qc.h(1) c) qc.x(0) qc.h0)
```

9.16 Which coding snippet will create a quantum circuit with three quantum bits and three classical bits?

```
a) QuantumCircuit(3,3) b) QuantumCircuit(3) c) QuantumCircuit([3,3])
```

9.17 Which of the following multi qubit-gate represents the controlled-z gate?

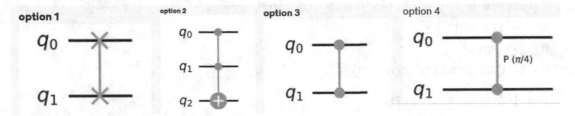

9.18 In the following quantum circuit, which statement should be added in the missing code to get the desired state $1/\mathrm{sqrt}(2)$ [1 0 0 0 0 0 0 1] (select 2).

```
qc = QuantumCircuit(3)
qc.h(0)
qc.cx([0,1],[1,2])
backend = BasicAer.get_backend('statevector_simulator')
job = execute(qc, backend, shots=1024)
result = job.result()
## Missing code

a) result.get_statevector(qc)
b) sv = StateVector.from_label('000')
```

```
    sv.evolve(qc).draw()
c) result.get_state(qc)
d) result.get_stateevolve(qc)
```

9.19 Which option will implement an operator that represents a single qubit Z-gate?

```
a) op = Operator(([1,0,0,1]]))
b) op = Operator(([j,0,0,-j]]))
c) qc = QuantumCircuit(2)
     qc.z(0)
     op = Operator(qc)
```

9.20 What is the output of the result variable in the following snippet?

```
q = QuantumRegister(1,'q')
qc = QuantumCircuit(q)
qc.y(0)
backend_unitary = BasicAer.get_backend('unitary_simulator')
result = execute(qc,backend_unitary).result().get_unitary(decimals=3)
```

```
a) ([0, -i],[i,0])
b) ([0, -i])
c) [0, -0.70i],[0.70i,0])
```

The algorithms in this section constitute the foundation upon which the more powerful Grover search and Shor factorization algorithms are built. Let's take a look at these remarkable creations in the next chapter.

Advanced Algorithms: Unstructured Search and Integer Factorization with Grover and Shor

This chapter brings two algorithms that have generated excitement about the possibilities of practical quantum computation:

- Grover's search: This is an unstructured quantum search algorithm created by Lov Grover which is capable of finding an input with high probability using a black box function or oracle. It can find an item in $O\left(\sqrt{N}\right)$ steps as opposed to a classical average of N/2 steps.

- Shor's integer factorization: The notorious quantum factorization that experts say could bring current asymmetric cryptography to its knees. Shor can factorize integers in approximately $\log(n^3)$ steps as opposed to the fastest classical algorithm, the Number Field Sieve at $\exp\left(k * \log\left(n^{\frac{1}{3}} \right) (\log\log n)^{\frac{2}{3}} \right)$.

Let's get started.

© Vladimir Silva 2024
V. Silva, *Quantum Computing by Practice*, https://doi.org/10.1007/978-1-4842-9991-3_10

Quantum Unstructured Search

Grover's algorithm is an unstructured search quantum procedure to find an entry of n bits on a digital haystack of N elements. As shown in Figure 10-1, Grover's quantum algorithm provides significant speedup at $O(\sqrt{N})$ steps. It may not seem much compared to the classical solution, but when we are talking millions of entries, then the square root of 10^6 is much faster than 10^6.

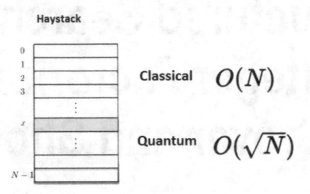

Figure 10-1. *Unstructured search time complexities*

If x is the element we are looking for, then Grover's algorithm can be described by the following pseudo code:

1. Prepare the input given f: {0, 1, ... , N-1} → {0,1}. Note that the size of the input is 2^n where n is the number of bits, and N is the number of steps or size of the haystack. The ultimate goal being find x such that f(x) = 1.

2. Apply a basis superposition to all qubits in the input.

3. Perform a phase inversion on the input qubits.

4. Perform an inversion about the mean on the input.

5. Repeat steps 3 and 4 at least $\sqrt{N}$ times. There is a high probability that x will be found at this point.

Let's take a closer look at the critical phase inversion and inversion about the mean steps.

Phase Inversion

This is the first step in the algorithm and must be performed in a superposition of all states in the haystack. If the element we are looking for is x' where $f(x') = 1$ then, the superposition can be expressed as $\sum \alpha |x\rangle$. Ultimately, what phase inversion does is:

$$\sum \alpha \; |x\rangle \underline{\text{Phase Inversion}} \begin{cases} \sum \alpha |x\rangle \; if \; x \neq x' \\ -\alpha |x'\rangle \; Otherwise \end{cases}$$

That is, if a given x is not the element we are looking for ($x \neq x'$), then it leaves the superposition intact; otherwise it inverts the phase (the sign of the complex coefficient α of the qubit – see Figure 10-2 for a pictorial representation).

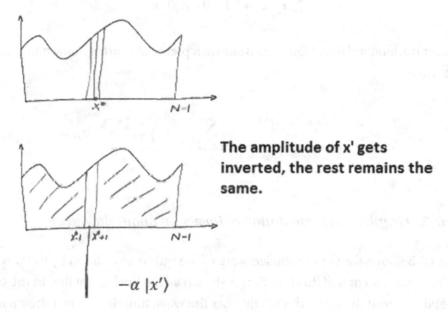

The amplitude of x' gets inverted, the rest remains the same.

Figure 10-2. *Pictorial representation of phase inversion*

This is the first step in Grover's algorithm; we'll see how phase inversion helps in finding the element we are looking for, but for now let's look at the second step: inversion about the mean.

Inversion About the Mean

Given the previous superposition $\sum \alpha | x \rangle$, we define the mean μ, as the average value of the amplitudes:

$$\mu = \frac{\sum_{x=0}^{N-1} \alpha_x}{N}$$

Now we must flip the amplitudes about this mean. That is:

$$\alpha_x \rightarrow \left(2\mu - \alpha_x \right)$$

$$\sum \alpha_x \ x \rightarrow \sum \left(2\mu - \alpha_x \right) \ x$$

To better understand this, Figure 10-3 shows a pictorial representation of inversion about the mean.

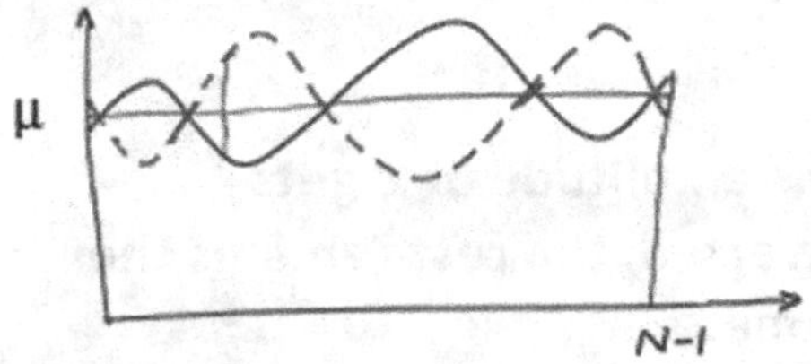

$$\sum \alpha_x |x\rangle \rightarrow \sum \left(2\mu - \alpha_x \right) |x\rangle$$

Figure 10-3. *Graphical representation of inversion about the mean*

Figure 10-3 shows the superimposed state of the qubits as defined by the wave function Ψ. The mean or μ of this function is shown as the horizontal line in the chart. What inversion about the mean does is mirrors the wave function Ψ over the mean μ resulting in a mirror wave (shown with a dotted line). This is equivalent to rotating the waves over the axis μ. Let's make sense of all this by putting all steps together to see them in action.

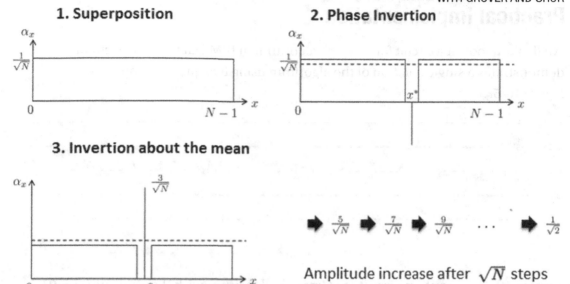

Figure 10-4. *Single Grover's iteration*

In Figure 10-4:

- The superposition of all qubits puts all amplitudes at $\frac{1}{\sqrt{N}}$.

- Next, a phase inversion puts the amplitude for x' at $-\frac{1}{\sqrt{N}}$. Note that this has the effect of lowering slightly the value of the mean μ, as shown by the dotted line in Figure 10-4 step 2.

- After the inversion about the mean, the mean amplitude drops a little bit, but x' goes way high, as much as $\frac{2}{\sqrt{N}}$ above the mean μ.

- If we repeat this sequence, the amplitude of x' increases by about $\frac{2}{\sqrt{N}}$ until that, in about $\sqrt{N}$ steps the amplitude becomes $\frac{1}{\sqrt{2}}$.

- At this point, if we measure our qubits, the probability of finding x' (the element we are looking for), as defined by quantum mechanics, is the square of the amplitude. That is ½.

- Thus, we are done. In roughly $\sqrt{N}$ steps, we have found the marked element x'.

Now, let's put all this together in a quantum circuit and corresponding code implementation.

Practical Implementation

We'll take a look at a circuit for Grover's algorithm in IBM Quantum. The circuit demonstrates a single iteration of the algorithm using two qubits as shown in Figure 10-5.

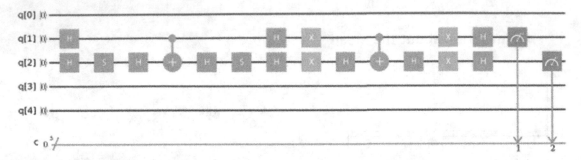

Figure 10-5. *Quantum circuit for Grover's algorithm with 2 qubits and A = 01*

A Python script that creates the circuit in Figure 10-5 is shown in Listing 10-1.

Listing 10-1. Python script for circuit in Figure 10-5

```
import sys,time,math
import numpy as np
from qiskit import *

# Import basic plotting tools
from qiskit.tools.visualization import plot_histogram

# Set the input bits to search for
def input_phase (circuit, qubits):
  # Uncomment for A = 00
  # Comment for A = 11
  #circuit.s(qubits[0])
  #circuit.s(qubits[1])
  return

# circuit: Grover 2-qubit circuit
# qubits: Array of 2 qubits
def invert_over_the_mean (circuit, qubits):
  for i in range (2):
```

```
    circuit.h(qubits[i])
    circuit.x(qubits[i])

  circuit.h(qubits[1])
  circuit.cx(qubits[0], qubits[1])
  circuit.h(qubits[1])

  for i in range (2):
    circuit.x(qubits[i])
    circuit.h(qubits[i])

def invert_phase (circuit, qubits):
  # Oracle
  circuit.h(qubits[1])
  circuit.cx(qubits[0], qubits[1])
  circuit.h(qubits[1])

def main():

  # Create qubits/registers
  size = 2
  q = QuantumRegister(size, 'q')
  c = ClassicalRegister(size, 'c')

  # Quantum circuit
  grover = QuantumCircuit(q, c, name = 'grover')

  #  loops = sqrt(2^n) * PI/4
  #loops = math.floor(math.sqrt(2**size) * (math.pi/4))

  # 1. put all qubits in superposition
  for i in range (size):
    grover.h(q[i])

  # Set the input
  input_phase(grover, q)

  # 2. Phase inversion
  invert_phase(grover, q)
```

```
input_phase(grover, q)

# 3. Invert over the mean
invert_over_the_mean (grover, q)

# measure
for i in range (size):
  grover.measure(q[i], c[i])

circuits = [grover]

# Execute the quantum circuits on the simulator
backend = Aer.get_backend("qasm_simulator")

result = execute(circuits, backend=backend).result()
counts = result.get_counts()
print("Counts:" + str(counts))

###########################################
# main
if __name__ == '__main__':
  start_time = time.time()
  main()
  print("--- %s seconds ---" % (time.time() - start_time))
```

- Listing 10-1 performs a single interaction of Grover's algorithm for a 2 bit input using two qubits. Even though the pseudo code in the previous section states that the total number of iterations is given by roughly $\sqrt{N}$ steps, the inversion about the mean requires this value to be multiplied by $\pi/4$ and its *floor* extracted (see the proof next to Figure 10-8). Therefore, we end up with $IT = floor\left(\sqrt{N} * \dfrac{\pi}{4}\right)$ where

  $N = 2^{bits}$. Thus, for two bits we get $IT = floor\left(\sqrt{4} * \dfrac{\pi}{4}\right) = floor(1.57) = 1$.

- The script begins by creating a quantum circuit with 2 qubits and two classical registers to store their measurements.

- Next, all qubits are put in superposition using the Hadamard gate.

- Before the iteration, the input is prepared using the phase gate (S) and the rules in Table 10-1.

320

Table 10-1. *Input preparation rules for Listing 10-1*

Input (A)	Gates/qubits
00	S(01)
10	S(0)
01	S(1)
11	None

- Next, perform a phase inversion followed by an inversion over the mean on the input qubits corresponding to a single iteration of the algorithm.

- Finally, measure the results and execute the circuit in the local or remote simulator. Print the result counts.

Generalized Circuit

In broad terms, the circuit in Figure 10-5 can be generalized to any number of input qubits as shown in Figure 10-6.

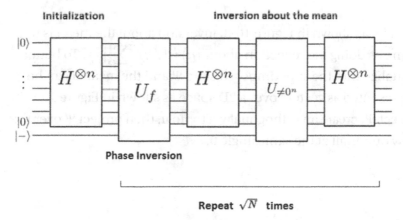

Figure 10-6. *Generalization of Grover's algorithm for an arbitrary number of qubits*

- The first box in Figure 10-6 puts all qubits in superposition by applying the Hadamard gate to the input of size n. This is the initialization step.

- Next, the phase inversion circuit (U_f) receives the superimposed input $\Psi = \sum \alpha \,|\, x \rangle$ and a phase input (minus state). This has the desired effect of putting the phase exactly where we want it. Thus, the output becomes $\sum \alpha \,(-1)^{f(x)} \,|\, x \rangle$. But how can this be achieved? The answer is that, by applying an exclusive OR on the minus state input, we obtain the desired effect $|b\rangle \rightarrow |\,f(x) \oplus b\rangle$ as shown in Figure 10-7. The third row of the XOR truth table between f(x) and b (the right side of Figure 10-7) shows the phase inversion effect.

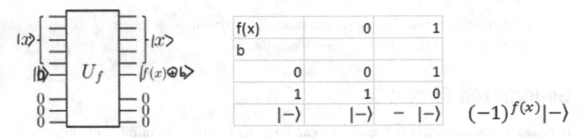

f(x)		0	1	
b				
0		0	1	
1		1	0	
$|-\rangle$		$|-\rangle$	$-\,|-\rangle$	$(-1)^{f(x)}|-\rangle$

Figure 10-7. *Phase inversion circuit*

- Finally, as shown in Figure 10-3, inversion about the mean is the same as doing the reflection about $|\mu\rangle = 1/\sqrt{N}\sum_{x}|x\rangle$. To better visualize this, the superimposed state Ψ and the mean μ can be represented as vectors over a 2D space as shown in Figure 10-8. To reflect Ψ, create an orthogonal vector to μ, then project Ψ over the new quadrant at the same angle θ.

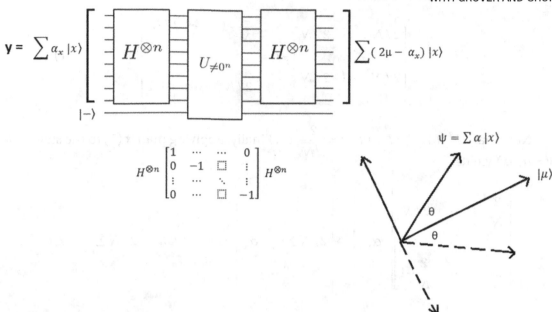

Figure 10-8. *Inversion over the mean circuit*

The proof that inversion over the mean transforms $\sum \alpha_x | x \rangle \rightarrow \sum (2\mu - \alpha_x) | x \rangle$ involves three steps, as shown by the circuit in Figure 10-8.

1. Transform $|\mu\rangle$ to the all zeros vector $|0, ..., 0\rangle$. This is achieved by applying the Hadamard gate to the input.

2. Reflect about the all zeros vector $|0, ..., 0\rangle$. This can be done by

     multiplying by the sparse matrix $\begin{bmatrix} 1 & \cdots & & \cdots & 0 \\ 0 & -1 & & & \vdots \\ \vdots & & \cdots & \ddots & \vdots \\ 0 & \cdots & & & -1 \end{bmatrix}$.

3. Transform $|0, ..., 0\rangle$ back to $|\mu\rangle$ by applying the Hadamard again. Thus:

$$H^{\otimes n} \begin{bmatrix} 1 & \cdots & & \cdots & 0 \\ 0 & -1 & & & \vdots \\ \vdots & & \cdots & \ddots & \vdots \\ 0 & \cdots & & & -1 \end{bmatrix} H^{\otimes n} = H^{\otimes n} \left(\begin{bmatrix} 2 & \cdots & 0 \\ \vdots & \ddots & \vdots \\ 0 & \cdots & 0 \end{bmatrix} - I \right) H^{\otimes n} = H^{\otimes n} \begin{bmatrix} 2 & \cdots & 0 \\ \vdots & \ddots & \vdots \\ 0 & \cdots & 0 \end{bmatrix} H^{\otimes n} - H^{\otimes n} I H^{\otimes n}$$

$$= \begin{bmatrix} 2/N & \cdots & 2/N \\ \vdots & \ddots & \vdots \\ 2/N & \cdots & 2/N \end{bmatrix} - I = \begin{bmatrix} \dfrac{2}{N}-1 & \cdots & 2/N \\ \vdots & \ddots & \vdots \\ 2/N & \cdots & \dfrac{2}{N}-1 \end{bmatrix} \tag{1}$$

Note that $H^{\otimes n} I\, H^{\otimes n} = I$ and $H = \dfrac{2}{\sqrt{N}}|x\rangle$. Finally, applying matrix (1) to the state $\Psi = \alpha_x \,|\, x\rangle$ yields:

$$\begin{bmatrix} \dfrac{2}{N}-1 & \cdots & 2/N \\ \vdots & \ddots & \vdots \\ 2/N & \cdots & \dfrac{2}{N}-1 \end{bmatrix} \begin{bmatrix} \alpha_0 \\ \vdots \\ \alpha_x \\ \vdots \\ \alpha_{N-1} \end{bmatrix} \rightarrow \begin{bmatrix} \vdots \\ 2/N \sum \alpha_y - \alpha_x \\ \vdots \end{bmatrix} = 2\mu - \alpha_x \; where \; 2/N \sum \alpha_y = 2\mu$$

So this is Grover's algorithm for unstructured search. It is fast, powerful, and soon to be hard at work on the data center cranking up all kinds of database searches. Given its significant performance boost over its classical cousin, chances are that in a few years, when quantum computers become more business friendly, most web searches will be performed by this quantum powerhouse. Before we finish it is worth noting, that by the time of this writing, a useful implementation or experiment (one that can find a real thing) does not exist for IBM Quantum. Hopefully, this will change in the future, but for now, Grover's algorithm lives in the theoretical side of things. In the next section, we close strongly by looking at the famous Shor algorithm for integer factorization.

Integer Factorization with Shor's Algorithm

The game of cat and mouse between cryptography and crypto-analysis rages on: the first, devising new ways to encrypt our everyday data; the latter probing for weaknesses, always looking for a crack to bring it down. Current asymmetric cryptography relies on the well-known difficulty of factoring very large primes (in the hundreds of digits range). This section looks at the inner workings of Shor's algorithm, a method that gives exponential speedup for integer factorization using a quantum computer. This is followed by an implementation using a library called ProjectQ. Next, we simulate for sample integers and evaluate the results. Finally we look at current and future directions of integer factorization in quantum systems. Let's get started.

Challenging Asymmetric Cryptography with Quantum Factorization

In the pivotal paper "Polynomial-Time Algorithms for Prime Factorization and Discrete Logarithms on a Quantum Computer,"[1] Peter Shor proposed a quantum factorization method using a principle known to mathematicians for a long time: find the period (also known as order) of an element a in the multiplicative group modulo N; that is, the least positive integer such that:

$$x^r \equiv 1 (\bmod N)$$

where N is the number to factor and r is the period of x modulo N.

Tip   Large integer factorization is a problem that has puzzled mathematicians for millennia. In 1976, G. L. Miller postulated that using randomization, factorization can be reduced to finding the period of an element A modulo N, thus greatly simplifying this puzzle. This is the basic idea behind Shor's algorithm.

Shor divided his algorithm in three stages, two of which are performed by a classical computer in polynomial time:

1.   Input preparation: Done in a classical computer in polynomial time log (n).

2.   Find the period r of the element a such that $a^r \equiv 1$ $(mod N)$ via a quantum circuit. According to Shor, this takes O((log n)2(log log n)(log log log n)) steps on a quantum computer.

3.   Post processing: Done in a classical computer in polynomial time log (n).

But why is there so much excitement about this method? Compare its time complexity (Big-O) against the current classical champ: The Number Field Sieve as shown in Table 10-2 (plus another fan favorite, the venerable quadratic sieve).

---

[1] Peter Shor. Polynomial-Time Algorithms for Prime Factorization and Discrete Logarithms on a Quantum Computer

Table 10-2. *Time complexities for common factorization algorithms*

Algorithm	Time complexity
Shor's	$(\log n)^2 (\log \log n)(\log \log \log n)$
Number Field Sieve	$\exp\left(c\left(\log n\right)^{1/3} \left(\log \log n\right)^{\frac{2}{3}}\right)$
Quadratic Sieve	$\exp\left(\sqrt{\ln n \ln \ln n}\right)$

Incredibly, Shor's algorithm has a polynomial time complexity, far superior to the exponential time by the Number Field Sieve, the fastest known method for factorization in a classical computer. As a matter of fact, experts have estimated that Shor's could factor a 200+ digit integer in a matter of minutes. Such a feat would rock the foundation of current asymmetric cryptography used to generate the encryption keys for all of our web communications.

Tip   Symmetric cryptography is highly resistant to quantum computation and thus to Shor's algorithm.

But don't panic yet; a practical implementation in a real quantum computer is still a long way off. Nevertheless, the algorithm can be simulated in a classical system using the slick Python library: Project Q. We'll run Project Q's implementation in a further section, but next let's see how period finding can solve the factorization problem efficiently.

Period Finding

Period finding is the basic building block of Shor's algorithm. By using modular arithmetic, the problem is reduced to finding the period (r) of the function
$f(x) = a^x \bmod N$ (see Figure 10-9).

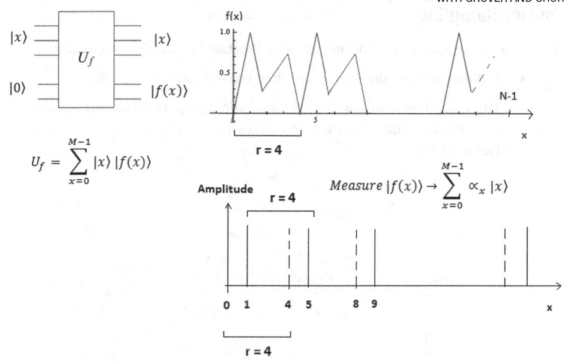

$$U_f = \sum_{x=0}^{M-1} |x\rangle \, |f(x)\rangle$$

$$Measure \; |f(x)\rangle \rightarrow \sum_{x=0}^{M-1} \propto_x |x\rangle$$

Figure 10-9. *Periodic function f(x)*

Figure 10-9 gives an example of a periodic function f(x) with period r = 4. For the algorithm to work, f(x) must meet three conditions:

1. f(x) is one-to-one on each period, that is, the values of f(x) must not repeat. In Figure 10-9 these values are represented by the vertices of each line per period.

2. For any given M or the number of periods, r must divide M. For example, given M = 100 and the period r = 4 M/r = 25.

3. M divided by r must be greater than r. That is $M > r^2$.

Shor's algorithm transforms f(x) into a quantum circuit U_f where the inputs are in superposition. If we measure the second register in U_f, we may see values for the amplitudes $\sum_{x=0}^{M-1} \propto_x x$ as shown in the amplitude chart of Figure 10-9. Here the amplitudes are exactly 4 units apart which is the period we are looking for. In this particular case, we get periodic superpositions with r = 4. But what do we do with this periodic superposition? Shor relies on another trick: Fourier sampling or Quantum Fourier Transform.

Fourier Sampling

Fourier sampling is a data manipulation process that has the following properties:

- It allows for input shifting without changing the output distribution.

- This is good because now we have a periodic superposition where the non-zero amplitudes are the multiples of the period (see Figure 10-10).

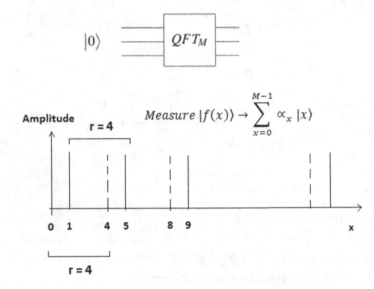

Figure 10-10. *Fourier sampling showing periodic superposition*

But what is the output of Fourier sampling? And how does it help? The answer is that its output is a random multiple of M/r. In this case, given M= 100 and r = 4, we get a random multiple of 100/4 = 25. This is advantageous for our goal. Let's see how.

Feed the Fourier Sampling Results to Euclid Greatest Common Divisor

If we were to run Fourier sampling multiple times, we will get random multiples of M/r. For example, we may get 50, 75, 25, etc. Now, if we apply Euclid's Greatest Common Divisor (gcd) to our random outputs, then viola: by dividing M by the gcd, we get the period r. Thus:

$$r = M/\gcd(50, 75, \ldots) = 100 / 25 = 4$$

So this is the outline for period finding via a quantum circuit. To understand how this method can find a factor efficiently, let's run through an example by factoring the number N = 21. Our task relies in two very efficient operations:

- Modular arithmetic: a = b (mod N). For example, 3 = 15 (mod 12).

- Greatest common divisor gcd(a, b). For example, gcd(15, 21) = 3.

Thus for N = 21, we need to solve the equation $x^2 \equiv 1 \; (mod\,21)$. That is, find the non-trivial square root x such that

- N divides (x +1) (x -1).

- N does not divide (x ± 1).

- Finally, recover a prime factor by applying gcd(N, x+1).

To find the non-trivial factor for N = 21, pick a random x. For example, given N = 21, choose x = 2, thus:

```
2⁰ ≡ 1  (mod 21)
2¹ ≡ 2  (mod 21)
2² ≡ 4  (mod 21)
2³ ≡ 8  (mod 21)
2⁴ ≡ 16 (mod 21)
2⁵ = 11 (mod 21)
2⁶ ≡ 1  (mod 21). Got the period r = 6.
```

In this case, $2^6 = (2^3)^2$. Thus, $2^3 = 8$ is a non-trivial factor such that 21 divides (8 + 1) (8 -1). Finally, we recover a factor with the greatest common divisor gcd (N, x+1) = gcd (21, 9) = 3. In general terms, pick an x at random, then loop through x^0, x^1,..., $x^r \equiv$ mod N. if we are lucky then r is even, that is, $(x^{r/2})^2 \equiv 1$ (mod N). And thus we have a non-trivial square root of 1 mod N.

Tip   It has been proven that, the probability that we get lucky, that is, r is even for $x^2 \equiv 1 \; (mod\,N)$ is ½. If we are unlucky, on the other hand, then we must repeat the procedure all over again. However, given the high probability of success, this would be insignificant in the great scheme of things.

Now, let's run the algorithm using the slick Python library ProjectQ.

Shor's Algorithm by ProjectQ

ProjectQ is an open source platform for quantum computing that implements Shor's algorithm using the circuit proposed by Stéphane Beauregard[2]. This circuit uses 2n + 3 qubits where n is the number of bits of the number N to factor. Beauregard's method is divided into the following steps:

1. If N is even, return the factor 2.

2. Classically determine if $N = p^q$ for $p \geq 1$ and $q \geq 2$, and if so return the factor p (in a classical computer this can be done in polynomial time).

3. Choose a random number a, such that $1 < a \leq N - 1$. Using Euclid's greatest common divisor, determine if gcd (a, N) > 1. If so, return the factor gcd(a,N).

4. Use the order-finding quantum circuit to find the order r of a modulo N. In a quantum computer, this step is done in polynomial time.

5. If r is odd or r is even but $a^{r/2} = -1 \pmod{N}$, then go to step (3). Otherwise, compute $gcd(a^{r/2} - 1 , N)$ and $gcd(a^{r/2} + 1 , N)$. Test to see if one of these is a non-trivial factor of N, and return the factor if so (in a classical computer this can be done in polynomial time).

---

[2] Stéphane Beauregard, Circuit for Shor's algorithm using 2n+3 qubits. Département de Physique et, Université de Montréal

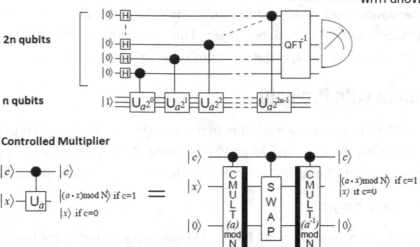

Figure 10-11. *Beauregard quantum circuit for period finding*

Beauregard implements period finding by using a series of controlled additions and multiplications in Fourier space to solve $f(x) = a^x(mod\ N) \rightarrow a^r \equiv 1\ mod\ N$ (see Figure 10-11):

- A controlled multiplier U_a maps $|x\rangle \rightarrow |ax\ (mod\ N)\rangle$ where

  - a is a classical relative prime to use as the base for a^x (mod N).

  - x is the quantum register.

  - c is the register of control qubits such that $U_a = ax$ (mod N) if c =1, and x otherwise.

- The controller multiplier U_a is in turn implemented as a series of doubly controlled modular adder gates such that

  - If both control qubits c1 = c2 = 1, the output is $f(x) = |\varphi(a + b\ mod\ N)\rangle$. That is, a + b (mod N) in Fourier space. Note that this gate adds two numbers: a relative prime (a) and a quantum number (b).

  - If either control qubit (c1, c2) is in state |0> then $f(x) = |\varphi(b)\rangle$.

- The doubly controlled modular adder gate is in turn built on top of the quantum addition circuit by Draper.[3] This circuit implements addition of a classical value (a) to the quantum value (b) in Fourier space.

Factorization with ProjectQ

Let's install ProjectQ and put the algorithm to the test. The first thing to do is to use the Python package manager to download and install ProjectQ (note that I am using Windows for the sake of simplicity. Linux users should be able to follow the same procedure):

```
C:\> pip install projectq
```

Next, grab the shor.py script from Project Q's examples folder[4] or the book source under Workspace\Ch10\p10-shor.py. Now, run the script and enter a number to factor (see Listing 10-2).

Listing 10-2. Shor's algorithm by ProjectQ in action. Keep in mind that the odds of success are not perfect; therefore, you may need to run a few times to obtain the correct factors

```
C:\>python shor.py
Number to factor: 21

Factoring N = 21: .........

Factors found : 7 * 3 = 21
Gate class counts:
    AllocateQubitGate : 166
    CCR : 1467
    CR : 7180
    CSwapGate : 50
    CXGate : 200
    DeallocateQubitGate : 166
    HGate : 2600
```

---

[3] T. Draper (2000), Addition on a quantum computer, quant-ph/0008033. Available online at https://arxiv.org/abs/quant-ph/0008033

[4] ProjectQ - An open source software framework for quantum computing. Available online at https://github.com/ProjectQ-Framework/ProjectQ

```
MeasureGate : 11
R : 608
XGate : 206
```

Gate counts:

```
Allocate : 166
CCR(0.098174770425) : 18
CCR(0.196349540849) : 30
CCR(0.392699081699) : 70
CCR(0.490873852124) : 18
CCR(0.785398163397) : 80
CCR(0.981747704246) : 38
CCR(1.079922474671) : 20
CCR(1.178097245096) : 16
   ...
R(5.252350217719) : 1
R(5.301437602932) : 1
R(5.497787143782) : 1
X : 206
```

```
Max. width (number of qubits) : 13.
--- 5.834410190582275 seconds ---
```

For N = 21, the script dumps a set of very helpful statistics such as

- The number of qubits used. Given N = 21 we need 5 bits, thus total-qubits = 2 * 5 + 3 = 13.

- The total number of gates used by type: In this case, doubly controlled CCR = 1467, CR = 7180, CSwap = 50, CX = 200, R = 608, X = 206, and others, for a grand total of 12646 quantum gates.

ProjectQ implements quantum period finding using Beauregard algorithm as shown in Listing 10-3:

- The run_shor function takes three arguments:

    - The quantum engine or simulator provided by project Q plus

    - N: the number to factor

    - a: The relative prime to use as a base for a^x mod N

- The function then loops from a = 0 to a = ln(N) with the quantum input register x in superposition; it then performs the quantum circuit for f(a) = a^x mod N as shown in Figure 10-11.

- Next, it performs Fourier sampling on the x register conditioned on previous outcomes and performs measurements.

- Finally, it sums the measured values into a number in range [0,1]. It then uses continued fraction expansion to return the denominator or potential period (r).

Listing 10-3. ProjectQ period finding quantum subroutine

```
def run_shor(eng, N, a):
  n = int(math.ceil(math.log(N, 2)))

  x = eng.allocate_qureg(n)

  X | x[0]

  measurements = [0] * (2 * n)  # will hold the 2n measurement results

  ctrl_qubit = eng.allocate_qubit()

  for k in range(2 * n):
    current_a = pow(a, 1 << (2 * n - 1 - k), N)

      # one iteration of 1-qubit QPE
    H | ctrl_qubit

    with Control(eng, ctrl_qubit):
      MultiplyByConstantModN(current_a, N) | x

    # perform inverse QFT --> Rotations conditioned on previous outcomes
    for i in range(k):
      if measurements[i]:
        R(-math.pi/(1 << (k - i))) | ctrl_qubit

    H | ctrl_qubit
```

```python
# and measure
All(Measure) | ctrl_qubit
eng.flush()
measurements[k] = int(ctrl_qubit)
if measurements[k]:
  X | ctrl_qubit

All(Measure) | x
# turn the measured values into a number in [0,1)
y = sum([(measurements[2 * n - 1 - i]*1. / (1 << (i + 1)))
     for i in range(2 * n)])

# continued fraction expansion to get the period
r = Fraction(y).limit_denominator(N-1).denominator

# return the (potential) period
return r
```

The next section compiles a set of factorization results for various values of N.

Simulation Results

ProjectQ's period finding subroutine is a simulation of a quantum circuit on a classical computer so it is not practical to use it to factorize large numbers. As a matter of fact, it is not capable to factor numbers larger than 4 digits in a reasonable time on a home PC. Table 10-3 shows a set of results for various values of N gathered from my laptop up to 2491.

Table 10-3. *Factorization results for various values of N*

Number (N)	Qubits	Time (s)	Memory (MB)	Quantum gate counts
15	11	2.44	50	CCR =792 CR =3186 CSwap = 32 CX = 128 H = 1408 R =320 X = 130 Measure = 9
105	17	27.74	200	CCR =3735 CR =25062 CSwap = 98 CX = 392 H = 6666 R =1568 X = 393 Measure = 15
1150	25	17542.12 (4.8h)	500	CCR =15366 CR =139382 CSwap = 242 CX = 968 H = 24222 R =5829 X = 981 Measure = 23
2491	27	246164.74 (68.3h)	2048	CCR = 20601 CR = 194670 CSwap = 288 CX = 1152 H = 31126 R =7509 X = 1166 Measure = 25

Factoring the 4-digit number 2491 took more than 68 hours on a 64 bit Windows 7 PC with an Intel® Core i-5 CPU @ 2.6 GHz with 16 GB of RAM. I tried to go a bit higher by attempting to factorize N = 8122 but gave up after 1 week. All in all, these results show that the algorithm can be simulated successfully for small numbers of N; however, it needs to be implemented in a real quantum computer to test its real power.

In this chapter, you have seen two algorithms that have generated excitement about the possibilities of practical quantum computation: Grover's algorithm, an unstructured quantum search method capable of finding inputs at an average of square root of N steps. This is much faster than the best classical solution at an average of N/2 steps. Expect all web searches to be performed by Grover's algorithm in the future.

Shor's algorithm for factorization in a quantum computer which experts say could bring current asymmetric cryptography to its knees. Shor's, arguably the most famous quantum algorithm out there, is a prime example of the power of quantum computation by providing exponential speedups over the best classical solution.

## CHAPTER 11

Quantum in the Real World: Advanced Chemistry and Protein Folding

My physics teacher used to say that quantum computers are notoriously bad calculators, and then he'll quote Richard Feynman to emphasize that these machines were conceived with atomic principles in mind; therefore, they should tackle problems at the atomic scale. Nowhere else is this more tangible than in the fields of chemistry and medicine where quantum is already working hard to make a difference. In this chapter, we showcase two amazing real-life experiments that illustrate how the power of quantum computation can make a difference in the real world.

The Significance of Eigenvalues

Eigenvalues take center stage in this context. An eigenvalue is a special scalar associated with a linear system of equations (matrix equation) extremely important in physics and engineering. The eigenvalue λ (lambda) is calculated from a transformation matrix (A) and vector (v) using the equation:

$$Av = \lambda v \ (11.1)$$

© Vladimir Silva 2024
V. Silva, *Quantum Computing by Practice*, https://doi.org/10.1007/978-1-4842-9991-3_11

The vector v is known as the eigenvector; now solving the equation for λ, we obtain:

$$(A - \lambda I)v = 0 \,(11.2)$$

where (I) is the identity matrix; note that it is mathematically legal to multiply the scalar λ by the identity matrix so it can be subtracted from matrix A. Finally, we can obtain λ by calculating the matrix determinant of:

$$\det(A - \lambda I) = 0 \,(11.3)$$

Here the eigenvector v cannot be zero; therefore, the determinant must be zero for equation 11.2 to hold.

Tip   Eigenvalues are also known as characteristic or latent roots and are heavily used in stability analysis, rotating body physics, oscillations of vibrating systems, molecular chemistry, graphics transformations, and many others.

Look at Figure 11-1 to understand this better; λ acts as a scalar of the transformation Av. Note that the direction could be opposite; however, the angle of the vector remains the same.

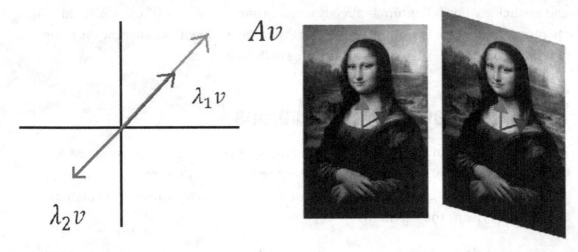

Figure 11-1. *Pictorial description on an eigenvalue along with a shear mapping of the Mona Lisa where the red vector is an eigenvector of the transformation*

Eigenvalues are important in

- Physics: They are used to measure the energy of a particle or the value of a measurable quantity associated with the wave function.

- Chemical engineering: They are used to solve differential equations and to analyze the stability of a system.

- Data science: They are used to extract the most important features of a dataset by identifying the directions of maximum variation in the data. These directions can be represented as eigenvectors, and the amount of variation can be represented as the eigenvalue.

Just to name a few, in our case, eigenvalues will be used to find ground states of a chemical element and in a protein folding experiment. Let's practice this important concept with some easy exercises.

Exercise 11.1: Find the eigenvalues and eigenvectors of the Pauli Matrix $X = \begin{bmatrix} 0 & 1 \\ 1 & 0 \end{bmatrix}$.

Tip: Use Equation 11.3 to find the 2 eigenvalues, and then use Equation 11.1 to find the eigenvectors.

$$\det\left(\begin{bmatrix} 0 & 1 \\ 1 & 0 \end{bmatrix} - \begin{bmatrix} \lambda & 0 \\ 0 & \lambda \end{bmatrix}\right) = 0, \ \det\left(\begin{bmatrix} -\lambda & 1 \\ 1 & -\lambda \end{bmatrix}\right) = 0, \lambda^2 = 1, \lambda = \pm 1$$

$$\text{for } \lambda = 1, \ \begin{bmatrix} 0 & 1 \\ 1 & 0 \end{bmatrix}\begin{bmatrix} x \\ y \end{bmatrix} = \begin{bmatrix} x \\ y \end{bmatrix}, \begin{bmatrix} y \\ x \end{bmatrix} = \begin{bmatrix} x \\ y \end{bmatrix}, \ v1 = \begin{bmatrix} 1 \\ 1 \end{bmatrix}$$

$$\text{for } \lambda = -1, \ \begin{bmatrix} 0 & 1 \\ 1 & 0 \end{bmatrix}\begin{bmatrix} x \\ y \end{bmatrix} = -1\begin{bmatrix} x \\ y \end{bmatrix}, \begin{bmatrix} y \\ x \end{bmatrix} = \begin{bmatrix} -x \\ -y \end{bmatrix}, \ v2 = \begin{bmatrix} 1 \\ -1 \end{bmatrix}$$

Try these on your own. As always, answers are provided in the appendix.

Exercise 11.2: Find the eigenvalues and eigenvectors of the Pauli Matrix $Y = \begin{bmatrix} 0 & -i \\ i & 0 \end{bmatrix}$.

Exercise 11.3: Find the eigenvalues and eigenvectors of the Pauli Matrix $Z = \begin{bmatrix} 1 & 0 \\ 0 & -1 \end{bmatrix}$.

What conclusion can you draw from the eigenvalues obtained from the XYZ Pauli matrices?

Eigenvalues in a Quantum Computer

To find eigenvalues in a quantum computer, we must use the Variational Quantum Eigensolver (VQE). This algorithm is made of the following components (see Figure 11-2):

- Ansatz: From the German for "guess," this is a parameterized quantum circuit for the wave function. As its name suggests, it is a conjecture where the parameters are a set of theta angles.

- Optimizer: This is a classical solver that takes the expected value of the wave function $\langle\Psi(\theta)|H|\Psi(\theta)\rangle$, evaluates its Hamiltonian (H), and adjusts the set of parameters θ to minimize the energy of the Hamiltonian.

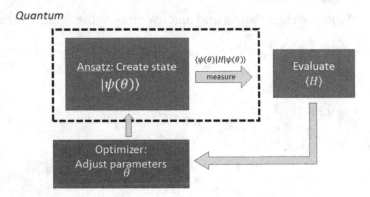

Figure 11-2. *Variational Quantum Eigensolver component layout*

This process repeats for a number of cycles (defined by the optimizer) until a desired minimum is reached. This would be the eigenvalue we are looking for.

Tip   In physics, a Hamiltonian is the total energy of a system, that is its kinetic energy (that of motion) and its potential energy (that of position).

But why use a Quantum computer if there are many classical solvers to do the job. Is there an advantage by using Quantum?

Why Use a Quantum Computer

Eigenvalues can be easily calculated in a classical computer; however, as the size of the transformation matrix (A) increases, any classical machine will run out of memory. Consider the estimated execution metrics for the classical Python and VQE solvers used to find the eigenvalues for a weighted lattice (see Table 11-1).

Table 11-1. *Execution metrics for the ground state calculation of the Kagome lattice using the Python's eigensolver vs. the VQE simulated and in-hardware[1]*

Lattice dimension	Execution time (s) Classic	Memory (bytes)	Execution time (s) quantum simulator	Execution time IBM-Quantum HW
12	2	64K	2	1.5h
13	12	192K	21	3.1h
14	116	520M	1104	?
15	1233	1G	7012	?
16	Crash	?	?	?

Note: Question marks indicate missing data due to lack of resources or hardware limitations.

Right now classical solvers are significantly faster than VQE (in simulation or hardware), however as the size of the transformation matrix increases, the execution time grows exponentially, and even worse the software runs out of memory and crashes. This is the fatal flaw of classical hardware. A quantum computer, on the other hand, has no memory and consumes qubit states in parallel. Currently, we live in the noisy quantum age, so the hardware is slow and limited to small samples; however, that will change in the future. Consider this: a 50 qubit transformation matrix requires 2^{50} = 1125 Terabytes of RAM in simulation. No supercomputer will ever be capable of going above 50 qubits, whereas a quantum computer can crunch thousands even millions of dimensions in parallel!

Thus VQE will be the algorithm that showcases this set of remarkable experiments already making a difference out there.

---

[1] High Fidelity Noise-Tolerant State Preparation of a Heisenberg spin-1/2 Hamiltonian for the Kagome Lattice on a 16 Qubit Quantum Computer. https://arxiv.org/abs/2304.04516

Molecule Ground States

Our goal is to compute the ground state (stationary state of lowest energy) of a lattice using a molecular Hamiltonian. Remember from the last section that the Hamiltonian of a system specifies its total kinetic and potential energy. In molecular chemistry, most elements are modeled using lattices where vertices represent interacting atoms; therefore, the minimum Hamiltonian will give us the value we are looking for. Thus, to reach the ground state, we need to minimize it.

Tip   Ground states are important as they tell where the excited electrons went to and returned from when they emit a photon. They also reveal the atomic number of elements. The ground state is known as the zero-point energy of the system.

The Lattice

When we think about the behavior of electrons in a solid, we can model them on a lattice where the vertices point to the position of atoms. This is a common practice in quantum physics, condensed matter physics, and quantum chemistry among others. Run Listing 11-1 to display a square lattice of six vertices (see Figure 11-3).

Note   Listing 11-1 uses custom modules not available in Qiskit or the Quantum Lab. Please run from the book source under Workspace\Ch11\vqelattice.py. The Lab is helpful for running quick and simple circuits; however, more complex programs with specialized dependencies can only be run from the command line.

Listing 11-1. Lattice construction

```
from qiskit_nature.second_q.hamiltonians.lattices import *
from qiskit_nature.second_q.hamiltonians import *
import matplotlib.pyplot as plt
from qiskit_nature.second_q.problems import LatticeModelProblem
from heisenberg_model import HeisenbergModel
from qiskit_nature.mappers.second_quantization import LogarithmicMapper
from qiskit.algorithms.minimum_eigensolvers import VQE
```

```
from qiskit.algorithms.optimizers import *
from qiskit.primitives import Estimator
from qiskit_nature.second_q.circuit.library import HartreeFock, UCCSD
from qiskit import *
from qiskit.primitives import BackendEstimator
from qiskit.circuit.library import *
from custom_vqe import CustomVQE

t = 1.0  # the interaction parameter
v = 0.0  # the onsite potential

rows=2
cols=3
nqubits = rows * cols

####### Square Lattice
square_lattice = SquareLattice(rows=rows, cols=cols, boundary_
condition=BoundaryCondition.PERIODIC)
square_lattice.draw()
#plt.savefig( 'sq.png')
plt.show()
```

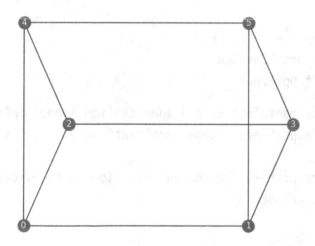

Figure 11-3. *Square lattice with six vertices and periodic boundaries*

> **Tip**   Qiskit implements an extensive library of lattice models for quantum chemistry. For details, see[2].

The Heisenberg Spin ½ Hamiltonian

We will use the Heisenberg Spin ½ model for the study of critical points and phase transitions of magnetic systems. It is given by the Hamiltonian:

$$H = -J\sum_{i,j}X_iY_iZ_i \otimes X_jY_jZ_j - h\sum_i X_iY_iZ_i$$

where $X = \begin{bmatrix} 0 & 1 \\ 1 & 0 \end{bmatrix}, Y = \begin{bmatrix} 0 & -i \\ i & 0 \end{bmatrix}, Z = \begin{bmatrix} 1 & 1 \\ 0 & -1 \end{bmatrix}$ are the Pauli spin-1/2 matrices, J is the coupling constant (the strength exerted in an interaction), and h is the external magnetic field. The spectrum of this Hamiltonian describes the statistical properties of a system in thermodynamic equilibrium (see Listing 11-2).

Listing 11-2.   Heisenberg model for the study of critical points and phase transitions of magnetic systems

```
"""The Heisenberg model"""
import logging
import numpy as np
from fractions import Fraction
from typing import Optional

from qiskit_nature.operators.second_quantization import SpinOp
from qiskit_nature.problems.second_quantization.lattice.lattices
import Lattice
from qiskit_nature.problems.second_quantization.lattice.models.lattice_
model import LatticeModel
```

---

[2] Qiskit Lattice models available at https://qiskit.org/ecosystem/nature/tutorials/10_lattice_models.html

```python
class HeisenbergModel(LatticeModel):
    """The Heisenberg model."""

    def coupling_matrix(self) -> np.ndarray:
        """Return the coupling matrix."""
        return self.interaction_matrix()

    @classmethod
    def uniform_parameters(
        cls,
        lattice: Lattice,
        uniform_interaction: complex,
        uniform_onsite_potential: complex,
    ) -> "HeisenbergModel":
        return cls(
            cls._generate_lattice_from_uniform_parameters(
                lattice, uniform_interaction, uniform_onsite_potential
            )
        )

    @classmethod
    def from_parameters(
        cls,
        interaction_matrix: np.ndarray,
    ) -> "HeisenbergModel":
        return cls(cls._generate_lattice_from_parameters(interaction_matrix))

    def second_q_ops(self, display_format: Optional[str] = None) -> SpinOp:
        if display_format is not None:
            logger.warning(
                "Spin operators do not support display-format. Provided
                display-format "
                "parameter will be ignored."
            )
        ham = []
        weighted_edge_list = self._lattice.weighted_edge_list
        register_length = self._lattice.num_nodes
```

```
# kinetic terms
for node_a, node_b, weight in weighted_edge_list:
  if node_a == node_b:
    index = node_a
    ham.append((f"X_{index}", weight))

  else:
    index_left = node_a
    index_right = node_b
    coupling_parameter = weight
    ham.append((f"X_{index_left} X_{index_right}", coupling_parameter))
    ham.append((f"Y_{index_left} Y_{index_right}", coupling_parameter))
    ham.append((f"Z_{index_left} Z_{index_right}", coupling_parameter))
  return SpinOp(ham, spin=Fraction(1, 2), register_
  length=register_length)
```

Now let's run the Heisenberg model within the main program to see how the Hamiltonian is constructed (see Listings 11-3, 11-4). Note that Listing 11-3 is a continuation of Listing 11-1.

Listing 11-3. Hamiltonian for the square lattice from Listing 11-1

```
# Build Hamiltonian from graph edges
heis = HeisenbergModel.uniform_parameters(
    lattice=square_lattice,
    uniform_interaction=t,
    uniform_onsite_potential=v,
)

# Map from SpinOp to qubits just as before.
log_mapper = LogarithmicMapper()
ham = 4 * log_mapper.map(heis.second_q_ops().simplify())
print(ham)
```

Listing 11-4. Heisenberg Hamiltonian for the square lattice using 6 qubits

```
1.0 * ZZIIII
+ 1.0 * ZIZIII
+ 1.0 * IZIZII
+ 1.0 * IIZZII
+ 1.0 * ZIIIZI
+ 1.0 * IIZIZI
+ 1.0 * IZIIIZ
+ 1.0 * IIIZIZ
+ 1.0 * IIIIZZ
+ 1.0 * YYIIII
+ 1.0 * YIYIII
+ 1.0 * IYIYII
+ 1.0 * IIYYII
+ 1.0 * YIIIYI
+ 1.0 * IIYIYI
+ 1.0 * IYIIIY
+ 1.0 * IIIYIY
+ 1.0 * IIIIYY
+ 1.0 * XXIIII
+ 1.0 * XIXIII
+ 1.0 * IXIXII
+ 1.0 * IIXXII
+ 1.0 * XIIIXI
+ 1.0 * IIXIXI
+ 1.0 * IXIIIX
+ 1.0 * IIIXIX
+ 1.0 * IIIIXX
```

Look at Listing 11-4 and Figure 11-3; the lattice has 9 edges, and our Hamiltonian is made of interactions of the X,Y,Z Pauli matrices; therefore, we get 27 (9*3) terms. Note that the indices in the Hamiltonian match the indices of the edges in Figure 11-3 (verify this by comparing the indices in the terms from Listing 11-4 with the indices from the connecting edges in Figure 11-3). With this in mind, we are ready to feed these objects to the VQE to extract the ground state of the lattice. Let's see how.

The VQE

The final piece of the puzzle is a *CustomVQE* object class (see Listing 11-5). It uses the minimum Eigensolver interface to compute a minimum eigenvalue for a given operator or Hamiltonian. Its job is to

Implement the objective function from Figure 11-2.

- Within the objective function, it invokes the Qiskit Runtime Estimator primitive service to obtain the expectation value of the quantum circuit (Ansatz) and observables (the terms in the Hamiltonian). Note that this value is the energy collected for each interaction of the optimizer and optionally sent back to the client.

- It runs the optimizer to minimize the energy. The optimizer runs for a number of cycles which is implementation specific. For each cycle, an energy value is produced and sent back through a result callback.

- It defines the initial point of random parameters (angles) for the Ansatz sent to the optimizer.

- It collects the final VQE energy result (after the optimizer completes) and returns it to the client.

Listing 11-5. Custom VQE for the ground state calculation

```
from qiskit.algorithms import MinimumEigensolver, VQEResult
import numpy as np

# Define a custom VQE class
class CustomVQE(MinimumEigensolver):

  def __init__(self, estimator, circuit, optimizer, callback=None):
    self._estimator = estimator
    self._circuit = circuit
    self._optimizer = optimizer
    self._callback = callback

  def compute_minimum_eigenvalue(self, operators, aux_operators=None):
```

```
# Define objective function to classically minimize over
def objective(x):
    # Execute job with estimator primitive
    job = self._estimator.run([self._circuit], [operators], [x])

    # Get results from jobs
    est_result = job.result()

    # Get the measured energy value
    value = est_result.values[0]
    # Save result information using callback function
    if self._callback is not None:
        self._callback(value)
    return value

# Select an initial point for the ansatzs' parameters
x0 = np.pi/4 * np.random.rand(self._circuit.num_parameters)

# Run optimization
res = self._optimizer.minimize(objective, x0=x0)

# Populate VQE result
result = VQEResult()
result.cost_function_evals = res.nfev
result.eigenvalue = res.fun
result.optimal_parameters = res.x
return result
```

Now we are ready to run the experiment. Collect Listings 11-1, 11-3, and 11-6 in a single file. Note that Listings 11-2 and 11-5 live in separate files in the same location. All files are provided with the source code of this manuscript.

The Results

Listing 11-6 is the third and final part of the main program. It does a few important things:

- It calculates the ground state using the classical solver NumPyEigensolver. This is for comparison purposes only to have a baseline against the quantum result.

- It invokes the CustomVQE to minimize the lattice Hamiltonian collecting energies at each step of the optimizer. These energies will be plotted along with the baseline classical ground state for comparison.

- A relative error between the last energy from the VQE and the classical ground state is computed to give a final relative error percentage. This tells us the accuracy of the experiment. The lower the better.

Listing 11-6. VQE execution and results

```
########## Classic
## Compute ground state energy
from qiskit.algorithms import NumPyEigensolver

# find the first three (k=3) eigenvalues
exact_solver = NumPyEigensolver(k=3)
exact_result = exact_solver.compute_eigenvalues(hamiltonian)
print(exact_result.eigenvalues)

# Save ground state energy for later
gs_energy = np.round(exact_result.eigenvalues[0], 4)
print("Ground state energy " + str(gs_energy))

######### VQE
ansatz = EfficientSU2(nqubits)
optimizer = NFT(maxiter=100)

#ansatz = TwoLocal(nqubits)
#optimizer = SLSQP(maxiter=75)
```

```python
intermediate_info = []
def callback(value):
  intermediate_info.append(value)

# Define instance of qiskit-terra's Estimator primitive
estimator   = Estimator() #[ansatz], [ham])

# Setup VQE algorithm
custom_vqe  = CustomVQE(estimator, ansatz, optimizer, callback=callback)

# Run the custom VQE function and monitor execution time
start       = time()
result      = custom_vqe.compute_minimum_eigenvalue(ham)
end         = time()

print(result)
print(f'execution time (s): {end - start:.2f}')

def rel_err(target, measured):
    return abs((target - measured) / target)

# Compute the relative error between the expected ground state energy
and the VQE
rel_error   = rel_err(gs_energy, result.eigenvalue)

print(f'Expected ground state energy: {gs_energy:.10f}')
print(f'Computed ground state energy: {result.eigenvalue:.10f}')
print(f'Relative error: {rel_error:.8f}')

#Let's plot the energy convergence data the callback function acquired.
plt.plot(intermediate_info, color='purple', lw=2, label='Simulated VQE')
plt.ylabel('Energy')
plt.xlabel('Iterations')
plt.axhline(y=gs_energy, color="tab:red", ls="--", lw=2, label="Target: " +
str(gs_energy))
plt.legend()
plt.grid()
plt.savefig("plot.png")
```

Run the experiment and compare your result with Figure 11-4 showing a 17% error between the VQE and classical solvers. Do you get a better result?

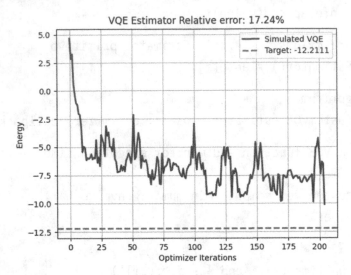

Figure 11-4. *VQE results from the ground state experiment*

Exercise 11.4: Change Listing 11-6 to use a different Ansatz, Optimizer combo from Qiskit circuit library[3] and Optimizer library[4]. Tip: All circuits are built using interfaces; thus, simply substitute the old names with new ones. Note that I have used (EfficientSU2, NFT): Compare your results with mine. See if you get a lower error rate.

Exercise 11.5: Qiskit packs an extensive library of lattices from [2]. Change Listing 11-1 to use other lattice types: For example, use the line lattice from the following snippet with a periodic boundary.

```
# Line
line_lattice = LineLattice (num_nodes=nqubits,     boundary_
condition=BoundaryCondition.PERIODIC)
line_lattice.draw(style=LatticeDrawStyle(with_labels=True, font_color='w'))
```

---

[3] Qiskit circuit library available at https://qiskit.org/documentation/apidoc/circuit_library.html

[4] Qiskit optimizer library available at https://qiskit.org/documentation/stubs/qiskit.algorithms.optimizers.html

All in all, VQE is a powerful algorithm in the quantum arsenal. It is so flexible, that it can be used pretty much in any optimization task. Let's continue with another remarkable experiment to showcase its power. This time in bioinformatics: enter protein folding.

Protein Folding

Proteins are the fundamental building blocks that power all life on Earth. A protein is a chain of amino acids that folds into a compact shape (conformation); they start as a linear chain of amino acids or a random coil with an unstable 3D structure. Eventually, the amino acids interact to form a well-defined, folded protein; these chemical machines create life. Reliably predicting protein structures is extremely complicated and can change our understanding of nature. Quantum computers can help unlock this complexity by predicting protein structure which is crucial to its function.

Note   Protein folding is the holy grail of biology, because the amino acid sequence determines the 3D structure of the protein; failure to fold properly produces a toxic compound that may cause a number of diseases.

There are a few important concepts to understand in this problem:

- Peptide: it is a small chain of amino acids (organic compounds made of amino and carboxylic acids – there are more than 500 amino acids in nature).

- Polypeptide: A longer, continuous, and unbranched peptide chain linked by chemical or peptide bonds.

- Folding stages: There are four stages that determine the protein structure and control its function:

    - Primary structure: This is a linear sequence of amino-acid residues in the polypeptide chain.

    - Secondary structure: Results from folds of the primary structure into either alpha helices or beta-sheets.

- Tertiary structure: Results from the folding of multiple secondary structures into one another. This is the geometric shape of the protein (see left side of Figure 11-5).

- Quaternary structure: Results from multiple tertiary structures interacting with each other to give rise to a functional protein such as hemoglobin (see right side of Figure 11-5).

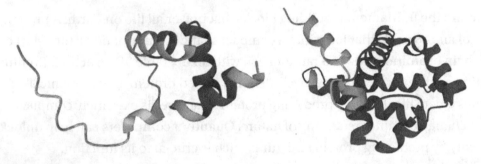

Figure 11-5. *Left E-coli protein showing primary, secondary, and tertiary structures. Right: hemoglobin showing quaternary structures (from the AlphaFold[5] Protein Structure Database)*

The Protein Folding Problem

According to the National Library of Medicine, the protein folding problem is made of three puzzles: (a) What is the folding code, (b) what is the folding mechanism, and (c) what is the protein 3D structure? Right now, only powerful supercomputers can be used to predict the 3D structure from its amino acid sequence along with sophisticated tools such as protein structure databases, computational physics methods, and complex algorithms like Monte Carlo sampling.[6]

---

[5] AlphaFold Protein Structure Database available online at https://alphafold.ebi.ac.uk/entry/A0A5E8GAP1

[6] The Protein Folding Problem from the National Library of Medicine available online at www.ncbi.nlm.nih.gov/pmc/articles/PMC2443096/

Tip   In 1994, CASP (Critical Assessment of Techniques for Protein Structure Prediction) was created to predict the unknown structures of proteins. CASP has grown from a small group into a worldwide initiative of more than 200 groups in 19+ countries.

Protein Folding Using a Quantum Computer

We can get into the protein folding game using Qiskit's research package, a library somewhat obscure by the time of this writing. This library is in alpha stage for now, and it is not part of the official distribution; however, it can be easily installed with the command:

```
pip install qiskit-research
```

Now we are ready to start folding. Let's see how.

Problem Initialization

The code in this section is divided into three parts. We start with the initialization step (see Listing 11-7). In this step we define

- A protein amino acid chain or peptide: In chemistry, amino acids are described using 1-letter notation. In this case the chain is APRLR: A = Alanine, P = Proline. R = Arginine, L = Leucine. For a full description of amino acid notation, see[7]. Note that a protein may contain side chains; these can bond with one another to hold the protein in a certain shape or conformation.

- Next, it defines an energy interaction process between the amino acids. In this case, we use the Miyazawa-Jernigan contact energy interaction. This is a widely used knowledge-based contact potential

---

[7] A One-Letter Notation for Amino Acid Sequences. https://febs.onlinelibrary.wiley.com/doi/pdf/10.1111/j.1432-1033.1968.tb00350.x

for globular proteins. It uses a probabilistic model to compare a one-body term with several hydrophobicity scales of amino acids. This method provides a strong correlation with layers of a protein when it is viewed as an ellipsoid[8].

- The contact energy interaction, along with the peptide is sent to the *ProteinFoldingProblem* which is an object that encapsulates this information to be passed to algorithms. Note that a set of penalty parameters is also defined. These parameters describe the strength of constraints enforcing the problem.

Note   Protein folding should be run from the command line via the book source at Workspace\Ch11\proteinfolding.py. This script cannot be run from the Quantum Lab.

Listing 11-7. Initialization step of the protein folding problem

```
from qiskit_research.protein_folding.interactions.miyazawa_jernigan_
interaction import *
from qiskit_research.protein_folding.peptide.peptide import Peptide
from qiskit_research.protein_folding.protein_folding_problem import *
from qiskit_research.protein_folding.penalty_parameters import
PenaltyParameters
from qiskit.utils import algorithm_globals, QuantumInstance
from qiskit.circuit.library import *
from qiskit.algorithms.optimizers import *
from qiskit.algorithms import NumPyMinimumEigensolver
from qiskit.algorithms.minimum_eigensolvers import SamplingVQE
from qiskit import execute, Aer
from qiskit.primitives import Sampler
import matplotlib.pyplot as plt
```

---

[8] The Miyazawa-Jernigan Contact Energies Revisited. https://openbioinformaticsjournal. com/contents/volumes/V6/TOBIOIJ-6-1/TOBIOIJ-6-1.pdf

```
# Protein without side chains
#main_chain = "APRLRFY"
#side_chains = [""] * 7

# with side chains
main_chain = "APRLR"
side_chains =["", "", "F", "Y", ""]

# interactions
#random_interaction = RandomInteraction()
mj_interaction = MiyazawaJerniganInteraction()

# phys constraints
penalty_back = 10
penalty_chiral = 10
penalty_1 = 10

penalty_terms = PenaltyParameters(penalty_chiral, penalty_back, penalty_1)

# peptide
peptide = Peptide(main_chain, side_chains)

# Problem
protein_folding_problem = ProteinFoldingProblem(peptide, mj_interaction,
penalty_terms)
qubit_op = protein_folding_problem.qubit_op()

# dump Hamiltonian
#print(qubit_op)
```

Running the VQE

To run the VQE, we need to define a few objects (see Listing 11-8):

- *Ansatz – RealAmplitudes*: It is a heuristic trial wave function used in chemistry applications or classification circuits in machine learning. The circuit is made of (RY) rotations over the y-axis of the Bloch sphere along with entanglements (CX gates) on neighboring qubits; its prepared quantum states will only have real amplitudes; the complex part is always 0.

- Optimizer – *COBYLA* (Constrained Optimization by Linear Approximation): It is a numerical optimization method for constrained problems where the derivative of the objective function is used to find the local minima.

- Hamiltonian: It describes the kinetic and potential energy of the system. This particular case uses a tetrahedral lattice (diamond shaped or cubic lattice) to encode the movement of amino acids through the configuration qubits. More details on how this encoding works are provided in[9].

Listing 11-8. VQE execution part for the protein folding problem

```
############## Run VQE
# classical optimizer
optimizer = COBYLA(maxiter=50)

# variational ansatz
ansatz = RealAmplitudes(reps=1)

counts = []
values = []

def store_intermediate_result(eval_count, parameters, mean, std):
    counts.append(eval_count)
    values.append(mean)

# initialize VQE using CVaR with alpha = 0.1
vqe = SamplingVQE(
    Sampler(),
    ansatz=ansatz,
    optimizer=optimizer,
```

---

[9] A.Robert, P.Barkoutsos, S.Woerner and I.Tavernelli, Resource-efficient quantum algorithm for protein folding, NPJ Quantum Information, 2021, https://doi.org/10.1038/s41534-021-00368-4

```
        aggregation=0.1,
        callback=store_intermediate_result,
)
raw_result = vqe.compute_minimum_eigenvalue(qubit_op)
```

Result Interpretation and Display

Finally, the binary probabilities returned by the VQE eigenstate are used to compute a best turn sequence bitstring which along with the peptide and unused qubits in the lattice are sent to a Protein Result class. This class interprets the bitstring and generates an XYZ file containing the Cartesian coordinates of each atom in the protein. This file is used to plot the protein in a 3D coordinate system (see Listing 11-9).

Listing 11-9. Result interpretation and display of the protein fold

```
########### Display Result interpretation
result = protein_folding_problem.interpret(raw_result=raw_result)
print(
    "The bitstring shape representation : ",
    result.turn_sequence,
)
print("The expanded expression:", result.get_result_binary_vector())
print(
    f"Main sequence of turns: {result.protein_shape_decoder.main_turns}"
)
print(f"Side turn sequences: {result.protein_shape_decoder.side_turns}")
print(result.protein_shape_file_gen.get_xyz_data())

########### Plot conformation (Shape) energy
fig = plt.figure()

plt.plot(counts, values)
plt.ylabel("Conformation Energy")
plt.xlabel("VQE Iterations")

fig.add_axes([0.44, 0.51, 0.44, 0.32])

plt.plot(counts[40:], values[40:])
```

```
plt.ylabel("Conformation Energy")
plt.xlabel("VQE Iterations")
plt.show()
#plt.savefig('protein-energy.png')

######### Plot Main chain
fig = result.get_figure(title="Protein Structure", ticks=False, grid=True)
fig.get_axes()[0].view_init(10, 70)
#plt.savefig('protein-main-chain.png')
plt.show()
```

Let's see the whole process in action:

Exercise 11.6: Concatenate Listings 11-7, 11-8, and 11-9, and run the experiment for Peptide (APRLRFY). Verify your results with Figure 11-6.

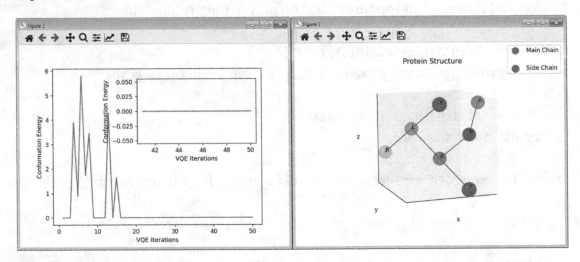

Figure 11-6. *Experimental results for the protein folding experiment using side chains*

Exercise 11.7: Alter your code to use the *EfficientSU2* Ansatz circuit which consists of layers of single qubit operations spanned by SU(2) (Special Unitary group of degree 2, or unitary matrices with determinant 1), and CX entanglements. This is a heuristic pattern to prepare trial wave functions for machine learning. Switch the optimizer to the Nakanishi-Fujii-Todo (NFT) algorithm (a method using gradient descent). Compare results for both experiments. Hint: rename the two instructions that create the Ansatz, Optimizer, and run; for a complete list of circuits and optimizers, see[10].

About the Peptide

The amino sequence used in this experiment is the neuro-peptide APRLRFY (NPID: NP02949) for organism: Aplysia californica (UniProt ID: ELH1_APLCA). You can get information about this protein from a neuro-peptide database such as[11]. The European Bioinformatics Institute also features a powerful 3D viewer to visualize your folds (see Figure 11-7).

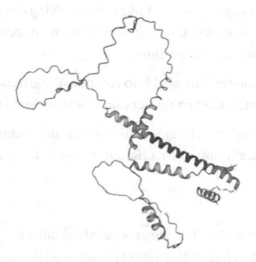

Figure 11-7. *3D visualization for neuro-peptide APRLRFY (NPID: NP02949) from the AlphaFold project from the European Bioinformatics Institute*

---

[10] Circuit Library available online at https://qiskit.org/documentation/apidoc/circuit_library.html

[11] Neuropeptide database - http://isyslab.info/NeuroPep/basic_search.jsp, AlphaFold Protein Structure Database -https://alphafold.ebi.ac.uk/

Exciting Times Lie Ahead

You have seen just a small sample of the impact quantum computers are having in the real world. But there is more, much more to cover for a single chapter. As a matter of fact, quantum is already carving inroads in major fields in science, industry, and business. Here is a small sample:

In physics and chemistry[12]

- Electronic structure: It aims to describe the quantum mechanical motion of a molecular system including the state of motion of electrons in an electrostatic field created by stationary nuclei.

- Vibrational structure: It is a technique for studying the structure and dynamics of photo excited molecules via monitoring of the vibrational spectrum in real-time.

- Ground state solvers: The goal is to compute the ground state of a molecular Hamiltonian using electronic or vibrational structure.

- Excited states solvers: The goal is to compute the excited states of a molecular Hamiltonian. This Hamiltonian can be electronic or vibrational.

In business[13]

- Portfolio optimization: It is the process of selecting the best asset distribution (portfolio), out of the set of all portfolios being considered.

- Portfolio diversification: It is the practice of spreading your investments around so that losses are limited.

- Credit risk analysis: It aims to determine the creditworthiness of a customer by assessing the probability that a customer will default on a payment before credit is extended.

- Option pricing theory: It provides an evaluation of an option's fair value, which traders incorporate into their strategies.

---

[12] Qiskit Nature available at https://qiskit.org/ecosystem/nature/tutorials/index.html

[13] Qiskit Finance available at https://qiskit.org/ecosystem/finance/tutorials/index.html

Quantum computers are poised to revolutionize information technology. By the end of 2023, IBM plans to release a 1000 qubit processor. A device that will be more powerful than all supercomputers ever created put together. Now is the time to learn about these incredible machines!

APPENDIX

Exercise Answers

Chapter 1

1. Lord Kelvin. 1900.

2. Max Planck. The ultraviolet catastrophe.

3. A *gnuplot* program:

```
set term jpeg
set termoption enhanced
set encoding utf8

# save as JPEG
set output 'uvc.jpg'

unset key
set tics nomirror out
set border 3
set xrange [0:3]
set yrange [0:15]

set ylabel "Spectral radiance" # (kW • sr-¹ • m-² • nm-¹)"
set xlabel "Wavelength (µm)"
set label "Planck 5000K" at 0.3,13.1
set label "4000 K" at 0.6,4.55
set label "3000 K" at 0.8,1.4
set label "Rayleigh-Jeans (5000K)" at 1.1,11
set grid lc rgb "light-blue"
```

© Vladimir Silva 2024
V. Silva, *Quantum Computing by Practice*, https://doi.org/10.1007/978-1-4842-9991-3

```
# length unit is micrometers
c=3e14 # speed of light
h=6.626e-22 # Planck constant
k=1.38e-11 # Boltzmann constant

# Planck curves at 3000, 4000, and 5000K
p1(x)=1e-6*2*h*c**2/(x**5*(exp(h*c/(x*k*3000))-1))
p2(x)=1e-6*2*h*c**2/(x**5*(exp(h*c/(x*k*4000))-1))
p3(x)=1e-6*2*h*c**2/(x**5*(exp(h*c/(x*k*5000))-1))

# Rayleigh-Jeans curve @ 5000K
rj(x)=1e-6*2*c*k*5000/(x**4)

plot p1(x) lw 2, p2(x) lw 2, p3(x) lw 2, rj(x) lw 2 lc rgb "black"
```

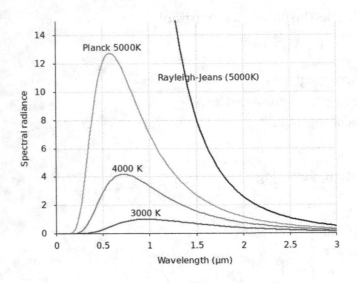

4. The photoelectric effect seeks to describe the behavior of
    electrons over a metal surface when light is thrown in the mix.
    Start with two metal plates (emitter and collector), attached via a
    cable to a battery. The negative end of the battery is connected to
    the emitter, and the positive to the collector. Measure the kinetic
    energy of the electrons when they flow from the emitter to the
    collector when a light source is thrown into the emitter; a vacuum
    must be set among the two. The photoelectric effect demonstrates
    the duality of the nature of light: both as particle (photon)
    and wave.

5. Albert Einstein.

    a. *Old notion:* If light flows as a wave as classical physics demands, then when the light hits the electrons, they will become energized and escape the surface of the emitter toward the collector. Furthermore, as the intensity (the amount) of light is increased, more electrons will get energized and escape in larger quantities.

    b. *Observed:* The increase in charge (the kinetic energy of the electrons) does not depend in the intensity of the light but in its frequency. Also, not all frequencies energize the electrons to escape the emitter. If we were to draw the kinetic energy (KE) as a function of the frequency (f), then there is a point in the curve (threshold frequency) after which the electrons escape. *Albert Einstein proposed a solution postulating that light behaves as a particle which he called a* **photon***.*

6. True or false: (a) true, (b) false, (c) false, (d) false, (e) false, (f) true, (g) true, (h) false, (i) true, (k) false.

7. Multiple choice: b, e, f.

8. We can measure the exact *position* or *momentum* of a particle but not both.

9. True.

10. At absolute zero atomic activity cannot seize because it will violate the uncertainty principle (momentum and position of particles will be known which is forbidden); therefore, there is still a tiny amount of energy.

11. Electrons cannot occupy the same quantum state at a time.

12. Multiple choice: false, true, true, true, false, true.

13. It was an attempt to account for special relativity (space-time coordinates) within Schrödinger's wave function.

14. Dirac equation describes the behavior of the electron in relativistic terms.

15. True: a, b, e.

16. Vacuum, positive, positron.

17. Entanglement is a fundamental property of quantum systems that originates when two particles interact with each other. For example, if one has spin-up, the other particle will instantaneously show spin-down when measured. This action propagates instantaneously across space, even time.

18. Albert Einstein. Because actions propagate instantaneously across space, time faster than the speed of light. No. Because quantum mechanics forbids information traveling faster than the speed of light.

19. Bell's theorem states that the sum of probabilities for a correlated three variable quantum system is less than or equal to 1. That is, $P(A = B) - P(A = C) - P(B = C) \leq 1$. It is important because it provides the means to test the principle of nonlocality in entangled particles.

20. Multiple choice: b, c.

21. Matter-wave, wave, particle, probabilistic, observation.

22. Many worlds is a deterministic interpretation: It eliminates the probabilistic nature of the Copenhagen interpretation.

    a. In many worlds there are no superposition of states; each history occurs in its own reality. In Copenhagen states of a particle exist in superposition until observed.

    b. In many worlds the wave function is universal. It does not collapse but it exists in infinite realities (worlds). In the Copenhagen interpretation, the collapse of the wave function signals the transition between the quantum and classical realms.

    c. In the double slit experiment on many worlds, all possible photon trajectories continue in their own timelines. In Copenhagen, photon trajectories are superimposed until the moment of observation (measurement).

23. Additional interpretations of quantum mechanics:

    a. Pilot-wave theory: This theory is deterministic (no randomness is allowed), no superposition of states, and accepts instantaneous action at a distance (nonlocality). It implies that configurations exist even when quantum systems are not observed. It uses a guiding equation which governs the evolution of the configurations over time.

    b. Quantum information: Postulates that notion that information is recorded irreversibly. The wave function evolves deterministically, going thru all possibilities. The conscious observer is allowed; however, it cannot gain knowledge until information has been recorded irreversibly. The measuring apparatus is quantum, but it can be statistically determined and capable or recording irreversible information.

24. Universe, vibrations, interact.

25. Electromagnetic field, lines of force.

26. Electron, neutrino, up quark, down quark. Gravity, electromagnetism, strong and weak nuclear forces.

27. The muon, the tau. All partners behave in the same way but have different mass.

28. Moun neutrino, tau neutrino, strange – bottom quarks, charm – top quarks.

29. The Higgs field is believed to be responsible for the mass of all particles in the standard model. It was discovered in the Large Hadron Collider (LHC) at 125 GeV.

30. The Lagrangian of the standard model is an equation to determine the state of a changing system and explain the maximum possible energy the system can maintain. It is important because it encompasses everything we have seen so far: the 12 particle fields, the 4 fields for the forces of nature, and the Higgs field.

Chapter 2

1. QFT is quantum field theory. QED is quantum electrodynamics. QED is a subset of QFT.

2. QED studies the interactions between the electron and electromagnetic fields.

3. Perturbation, theory, renormalization, and Feynman diagrams.

4. Electron scattering is the interaction of two electrons bumping into each other. In classical electro dynamics, this interaction is described by Coulomb's law.

5. $F = k\dfrac{q_1 q_2}{r^2}$ where k = Coulomb's, q1, q2 = charges of the electrons, and r = distance between them

6. True: a, b.

7. Equations. Virtual photon.

8. True: b, c, d.

9. Loop interactions are weird intermediate states that are problematic in Feynman diagrams because they may increase the effective mass of an electron arbitrarily. Examples:

   - A photon momentarily becomes an electron, positron pair, and reverts to a photon again.

   - An electron emits then reabsorbs the same photon.

10. Loop interactions become infinite because to calculate the mass correction from self-energy loops, we need to add all possible photon energies, but those energies can be arbitrarily large, thus sending the corrected mass to infinity.

11. Renormalization says: don't start with the corrected (or fundamental) mass of the electron but the experimentally measured mass instead. On other words, don't use its theoretically calculated value (infinite) but its experimental (finite) number, and then solve the equations from there.

12. In renormalization, for each infinity you want to get rid of, you must measure some property in the lab. It can't predict that particular property from scratch; it can only predict other properties relative to your lab measurement.

13. True: a.

14. Elements of a Feynman diagram:

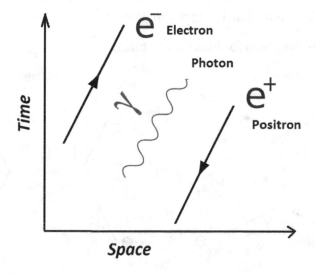

15. Most probable interaction for electron scattering:

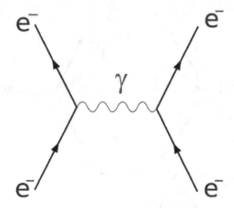

Two electrons (e⁻) move toward each other. They exchange a virtual photon (γ) and move away.

16. Virtual particles exist between vertices within the diagram but don't enter or leave. They are also by definition unmeasurable.

17. Two weird characteristics of virtual particles:

    • They do not obey the mass-energy-momentum equivalence and are dubbed *off-shell*.

    • They are not limited by the speed of light or the direction of time.

18. Feynman diagrams reduce the number of contributing interactions that need to be solved. They can easily represent paths backward in time using antimatter.

19. Two vertex diagrams for Bhabha scattering:

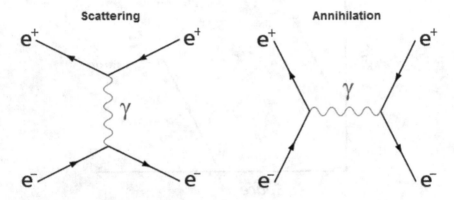

20. Four vertex diagrams for Bhabha scattering:

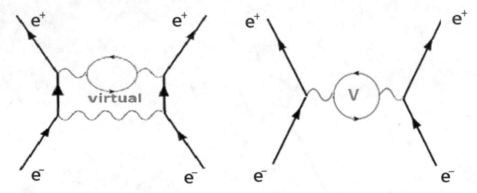

21. Continuous or reflective.

22. Two types of symmetries in nature

- Continuous: The concept of viewing symmetry as motions or changes in position over time. Circle divided in the middle.

- Discrete: It describes non-continuous (disconnected) changes in a system. Charge conjugation.

23. Spatial axis. Rotation.

24. When something remains identical under a parity transformation, it is said to be parity symmetric or P-symmetric. A full parity inversion involves flipping all three axes: X, Y, Z.

25. Flipped: Spatial axes: x, y, z. Not flipped: Angular momentum, energy, mass, gravity and time.

26. True.

27. Yes. The Wu experiment showed in 1956 that electrons produced in the decay if the Colbalt-60 radioactive isotope emerged in the opposite direction on the spin. This was the smoking gun that proved that parity symmetry is violated because of the correlation of spin and momentum of the ejected electron. Parity symmetry violations are the result of the weak atomic force.

28. CP symmetry involves charge conjugation and parity transformation. Experiments have indicated that CP symmetry is violated in our universe.

29. Cronin and Fitch sent a bunch of both types of Neutral Kaon (Hs, KL) down a tube into a detector in the far end. The short-lived Ks particles should never have completed the journey given their short lives, and yet, a significant number of decay products from Ks particles were found at the detector. The only explanation is that the long-lived KL particles transformed into Ks particles violating CP symmetry.

30. CPT symmetry means charge conjugation, parity inversion, and time reversal. The laws of physics must work the same under a flip of charge, parity, and direction of time (CPT transformation).

31. Two examples of T-symmetry transformations

- Time reversal is when we flip the so-called *arrow of time*, having the universe travel backward in time.

- Flipping the direction of the evolution of a physical system, that is, reverse all momentum and spin.

Chapter 3

1. $\sigma_x = \begin{bmatrix} 0 & 1 \\ 1 & 0 \end{bmatrix} \sigma_y = \begin{bmatrix} 0 & -i \\ i & 0 \end{bmatrix} \sigma_z = \begin{bmatrix} 1 & 0 \\ 0 & -1 \end{bmatrix}$

2. $\begin{bmatrix} 1 & 0 & 0 & 0 \\ 0 & 1 & 0 & 0 \\ 0 & 0 & 1 & 0 \\ 0 & 0 & 0 & -1 \end{bmatrix}$

3. Crystal cube constructed using two right triangle prisms. Linear plates made of a thin, flat glass plate that has been coated on the first surface of the substrate.

4. The high error probability.

5. When the state of a photon is not recorded then is not a measuring device, then we must add all the possible states (paths) to calculate its probability amplitude.

6. When two identical single photons enter a 1:1 beam splitter.

7. Constructive: add two waves. Destructive: waves cancel each other.

8. When waves cancel each other.

9. Reflectivity error for the horizontally polarized photons. Reflectivity error for the vertically polarized photons.

10. When electricity flows through a conductor with no resistance.

11. Mercury, Niobium-nitride.

12. 10mk.

13. Signal amplifier, mixing chamber, and cryoperm shield.

14. Cables used to raise the overall current carrying capability.

15. To provide thermal insulation while maintaining the best radiant energy barrier available.

16. To probe its readout pulse with a weak microwave signal (7Ghz).

17. Low Noise Amplifier (LNA), Quantum Limited Amplifier (QLA).

18. The probability of a successful determination of the qubit state.

19. Error level, longevity, and cost.

20. The probability of finding a particle at a given point is proportional to the square of the magnitude of the particle's wave-function at that point.

21. Ion traps, silicon quantum dots, diamond vacancies.

Chapter 4

3.
```
{
    "id": "access-token",
    "ttl": 1209600,
    "created": "2023-05-18T18:15:04.557Z",
    "userId": "5ae875060f0205003931559a"
}
```
5. [
```
    {
        "name": "ibm_lagos",
        "deleted": false,
        "costParameters": {
            "fixedOverhead": 20,
            "calCircuits": 0,
            "repRate": 3500
        },
```

```
            "specificConfiguration": {
                "simulator": false,
                "backend_name": "ibm_lagos",
                "backend_version": "1.2.5",
                "basis_gates": [
                    "cx",
                    "id",
                    "rz",
                    "sx",
                    "x"
                ],
                "max_shots": 100000,
                "max_experiments": 900,
            }
            ...
    ]
```

6.
```
    {
        "backend_name": "ibm_perth",
        "backend_version": "1.2.7",
        "gates": [
            {
                "gate": "id",
                "name": "id0",
                "parameters": [
        ...
    }
```

7.
```
    {
        "state": true,
        "status": "active",
        "message": "active",
        "lengthQueue": 586,
        "backend_version": "1.1.43"
    }
```

8.

```
[
    {
        "qasms": [
            {
                "qasm": "\ninclude \"qelib1.inc\";\nqreg q[4];
                \ncreg cr[4];\nu2(3.14159265358979,3.14159265358979) q[1];
                \nu3(-3.14159265358979,1.57079632679490,4.71238898038469) q[0];
                \ncx q[2],q[1];\nu2(0,3.14159265358979)            "ip": {
            "ip": "162.202.136.6",
            "country": "United States",
            "continent": "North America"
        },
        "id": "5bb19bca6975ec004aa5ab7b",
        "userId": "5ae875060f0205003931559a",
        "maxCredits": 3,
        "usedCredits": 0
    }
]
```

9.

```
{
    "total": 30,
    "count": 30,
    "codes": [
        {
            "id": "6330b11c0d4167a5ab08cc83",
            "description": {
            ...
            "userId": "5ae875060f0205003931559a",
            "lastUpdateDate": "2018-04-10T23:11:36.566Z"
        }
    ]
}
```

10.
```
{
    "api-q": "0.153.0",
    "api-app": "0.153.0",
    "api-utils": "0.153.0",
    "version": "0.153.1"
}
```

Chapter 5

5.4

$$v = \begin{bmatrix} 1.2 \\ 3.0 \\ -0.1 \end{bmatrix}$$

5.5

$$a.b = 1*4 + 2*5 + 3*6 = 32$$

5.7

$$A^T = \begin{bmatrix} 1 & 2 & 3 \end{bmatrix}$$

5.9

$$v^\dagger = \begin{bmatrix} a^* & b^* & c^* \end{bmatrix}$$

5.10

$$v^\dagger v = |a|^2 + |b|^2 + |c|^2$$

5.11

$$\begin{bmatrix} ax & ay & az \\ bx & by & bz \\ cx & cy & cz \end{bmatrix}$$

5.12

4i

5.13

$$i^{-3} = i, \, i^{-2} = -1, \, i^{-1} = -i, \, i^0 = 1, \, i^1 = I, \, i^2 = -1, \, i^3 = -i$$

5.15

$$\text{Cos } x = \tfrac{1}{2}(e^{ix} + e^{-ix}) = \tfrac{1}{2}[\text{Cos } x + i\text{Sin } x + \text{Cos}(-x) + i\text{Sim}(-x)]$$
$$= \tfrac{1}{2}[\text{Cos } x + i\text{Sin } x + \text{Cos } x - i\text{Sinx}]$$
$$= \tfrac{1}{2}[2\text{Cos } x] = \text{Cos } x$$

5.16

$$\text{Sin } x = \tfrac{1}{2}i\,(e^{ix} - e^{-ix}) = 1/2i\,[\text{Cos } x + i\text{Sin } x - [\text{Cos}(-x) + i\text{Sin}(-x)]]$$
$$= 1/2i\,[\text{Cos } x + i\text{Sin } x - \text{Cos } x + i\text{Sin } x]$$
$$= 1/2i\,[2i\text{Sin } x] = \text{Sin } x$$

5.19

$$|+\rangle \rightarrow \cos\left(\frac{\pi}{4}\right)|0\rangle + e^{i\pi/4}\sin\left(\frac{\pi}{4}\right)|1\rangle = 1/\sqrt{2}\left(|0\rangle + |1\rangle\right)$$

$$|-\rangle \rightarrow \cos\left(-\frac{\pi}{4}\right)|0\rangle + e^{i\pi}\sin\left(-\frac{\pi}{4}\right)|1\rangle = 1/\sqrt{2}\left(|0\rangle - |1\rangle\right)$$

5.20

$$|i\rangle \rightarrow \cos\left(\frac{\pi}{4}\right)|0\rangle + e^{i\pi/2}\sin\left(\frac{\pi}{4}\right)|1\rangle = 1/\sqrt{2}\left(|0\rangle + |i\rangle\right)$$

$$|-i\rangle \rightarrow \cos\left(\frac{\pi}{4}\right)|0\rangle + e^{i3\pi/4}\sin\left(\frac{\pi}{4}\right)|1\rangle = 1/\sqrt{2}\left(|0\rangle - |i\rangle\right)$$

5.21

$$H$$

5.22 Half Adder

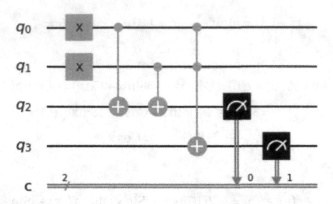

5.23

$$z^* = a - ib$$

5.24

$$a + c + i(b + d)$$

5.25

$$(ac - bd) + i(ad + bc)$$

5.26

$$z = \sqrt{Re(z)^2 + Im(z)^2}$$

5.27

$$\left[ax0 + by0 + cz0 \quad ax1 + by1 + cz1 \quad ax2 + by2 + cy2 \right]$$

5.28

$$X, H, Z$$

Gates that are their own unitary inverses are called Hermitian operators. Elementary gates such as the Hadamard (H) and the Pauli gates (I, X, Y, Z) are Hermitian operators, while others like the phase shift (S, T, P, CPHASE) gates are not.

5.29

$$\Pi$$

Chapter 6

6.1

```
from qiskit import *
from qiskit.tools.visualization import *

# create a 1 qubit  circuit with  1 classic register
qc = QuantumCircuit(1,1)

# Pauli X gate
qc.x(0)

# measure gate from qubit 0 to classical bit 0
qc.measure(0, 0)

# backend simulator
backend = 'qasm_simulator'

# run in simulator
job = execute(qc, Aer.get_backend(backend))
```

```
# Show result counts
print (job.result().get_counts())

# Print in studout
qc.draw()
```

   6.2

   6.3

```
plot_histogram(job.result().get_counts())
```

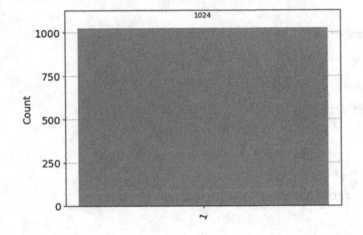

6.4

```
from qiskit import *
from qiskit.visualization import *

qc = QuantumCircuit(2)
qc.x(0)
qc.h(0)
qc.cx(0,1)
qc.measure_all()

job = execute(qc, Aer.get_backend
("qasm_simulator"))
plot_histogram (job.result().get_
counts())

from qiskit import *
from qiskit.visualization import *
qc = QuantumCircuit(2)
qc.x(0)
qc.h(0)
qc.cx(0,1)

job - execute(qc, Aer.get_
backend("statevector_simulator"))
plot_state_city(job.result().get_
statevector())
```

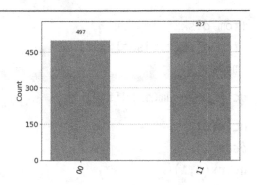

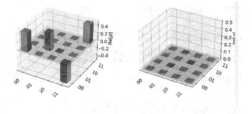

6.5

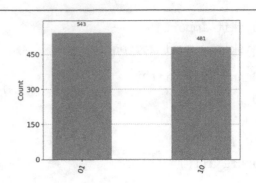

```
from qiskit import *
from qiskit.visualization
import *

qc = QuantumCircuit(2)

qc.h(0)
qc.x(1)
qc.cx(0,1)
qc.measure_all()

job = execute(qc, Aer.get_
backend("qasm_simulator"))
plot_histogram (job.result().
get_counts())
```

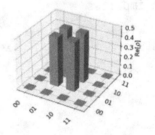
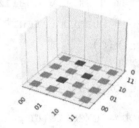

```
from qiskit import *
from qiskit.visualization
import *

qc = QuantumCircuit(2)
qc.h(0)
qc.x(1)

qc.cx(0,1)

job = execute(qc, Aer.
get_backend("statevector_
simulator"))
plot_state_city(job.result().
get_statevector())
```

6.6

```
from qiskit import *
from qiskit.tools.visualization import*
from qiskit.providers.fake_provider
import *

qc = QuantumCircuit(3)

qc.h(0)
qc.cx(0,1)
qc.cx(0,2)
qc.measure_all()

print (qc)
backend = FakePerth()

job = execute(qc, backend)
plot_histogram(job.result().get_
counts())
```

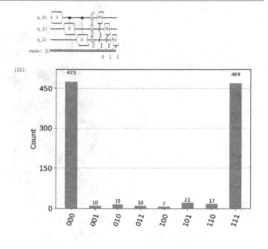

6.7

```
# create a 2 qubit  circuit with  1 classic register
qc = QuantumCircuit(2)

# Phi+
qc.h(0)
qc.cx(0,1)

qc.measure_all()

# backend simulator
backend = 'qasm_simulator'

# run in simulator
job = execute(qc, Aer.get_backend(backend))

# Show result counts
print (job.result().get_counts())

# Print in studout
qc.draw()
```

6.8

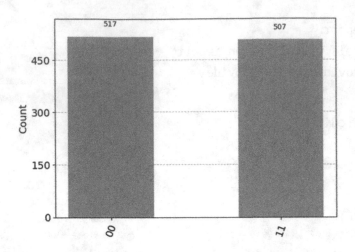

6.9

```
qc = QuantumCircuit(2)

# Phi-
qc.x(0)
qc.h(0)
qc.cx(0,1)

# backend simulator
backend = 'statevector_simulator'

# run in simulator
job = execute(qc, Aer.get_backend(backend))

# Print in stdout
array_to_latex(job.result().get_statevector())
```

6.10

```
# create a 2 qubit  circuit with  1 classic register
qc = QuantumCircuit(2)

# psi +
qc.h(0)
qc.x(1)
```

```
qc.cx(0,1)

qc.measure_all()

# backend simulator
backend = 'qasm_simulator'

# run in simulator
job = execute(qc, Aer.get_backend(backend))

# Show result counts
print (job.result().get_counts())

# Print in studout
qc.draw()
```

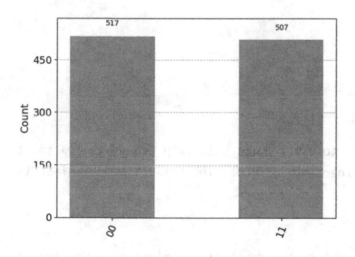

6.11

```
qc = QuantumCircuit(2)

# Psi-
qc.h(0)
qc.x(1)
qc.cx(0,1)
qc.z(1)

# backend simulator
backend = 'statevector_simulator'
```

```
# run in simulator
job = execute(qc, Aer.get_backend(backend))

# Print in studout
array_to_latex(job.result().get_statevector())
```

6.12

X

6.13

```
qc.qasm()
```

6.14

1

6.15

```
import qiskit
print(qiskit.__qiskit_version__)
```

6.16

Applying X-gate to qubit 0, it will flip the state |0> to |1>. Therefore the corresponding state is |01>. The default number of shots is 1024. {'01': 1024}

6.17

qc.depth()will return the depth of any given quantum circuit.

6.18

b) qc.y(1). Applying Y-gate to the qubit 1 will put in i|1> state and qubit 0 will be in |0> state. The resultant tensor-product will be i|10>.

6.19

S, T, I: Quantum gates such as S,T and I will leave the state |0> unchanged.

6.20

In the quantum circuits, with no classical register specified in the code, we have to use qc.measure_all() to measure the output of the qubit.

6.21

qc.measure([0,1],[0,1]). The measurement part to the circuit is considered as non-unitary.

6.22

a) For adding the legend and title use legend and title respectively in plot_histogram

```
legend = ['All H-gates']
title = "superposition states of 3 qubits"
plot_histogram(counts,legend=legend,title=title)
```

6.23

The Bell state returned by the given quantum circuit is $1/\sqrt{2}\,(|10> + |01>)$ and **option 1** suits best for that (with a positive phase).

option 1

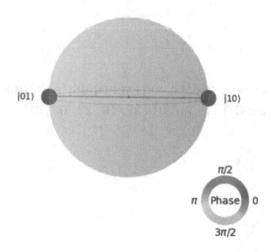

6.24

The bell circuit returns $1\sqrt{2}$ (|00> - |11>) which is plotted correctly in **option 1**.

option 1

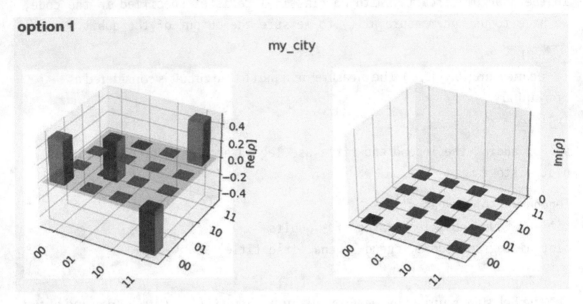

6.25

a,b,c. Note: the X gate has no effect when a qubit is in superposition.

Chapter 7

7.1

Applying Hadamard gate to each and every qubit in the quantum circuit will put the circuit in superposition: C) {'11': 30, '01': 27, '10': 22, '00': 21}.

7.2

{'000': 51, '111': 49}

7.3

1/sqrt(2) (|11> + |10>)

7.4

A, B, D

7.5

12

7.6

a) execute (qc, sim, shots=2000)

7.7

a) get_unitary()

7.8

memory (bool): If True, per-shot measurement bitstrings are returned as well (provided the backend supports it).

7.9

d) qc.draw(output='png') - PNG format is not supported.

7.10

b) Invalid: a) has an invalid simulator, c) has invalid parameter repeat.

Chapter 9

9.3

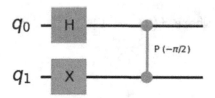

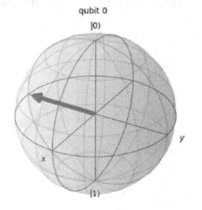

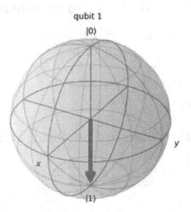

9.5

```python
from qiskit import QuantumRegister, ClassicalRegister, QuantumCircuit
from qiskit import Aer, execute
from qiskit.tools.visualization import plot_histogram

# Create quantum/classical registers and a quantum circuit
n = 3

qin     = QuantumRegister(n, 'x')
qout    = QuantumRegister(1, 'y')
c       = ClassicalRegister(n)
qc      = QuantumCircuit(qin,qout,c)

# This is balanced function
def balanced(n, circ, input, output) :
    for i in range (n):
        circ.cx(input[i],output)

# This is a constant function that doesn't change anything
def constant(circ, input, output) :
    circ.i(output)

# Build the Deutsh-Josza circuit

# First apply phase shift to negate where f(x) == 1
qc.h(qin)

qc.x(qout)
qc.h(qout)

#is equal2() for balanced or constant2() for constant in line below
#constant(qc,qin,qout)
balanced(n, qc,qin,qout)

qc.h(qout)
qc.barrier()

# Then Walsh-Hadamard on input bits
qc.h(qin)
```

```
# For constant function, output is |00> with probability 1
# For balanced function, output is something other than |00> with
probability 1
qc.measure(qin,c)

display(qc.draw())

# Simulate and show results
backend     = Aer.get_backend('qasm_simulator')
job         = execute(qc, backend, shots=512)  # shots default = 1024
result      = job.result()

plot_histogram(result.get_counts())
```

9.7

$$H^{\otimes 2}|01\rangle = \frac{1}{\sqrt{2}}\big(|0\rangle + |1\rangle\big) \otimes \frac{1}{\sqrt{2}}\big(|0\rangle - |1\rangle\big) = \frac{1}{2}\big(|00\rangle - |01\rangle + |01\rangle - |11\rangle\big)$$

9.9

011

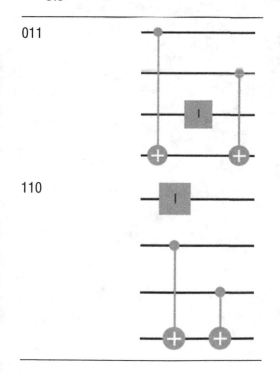

110

9.10

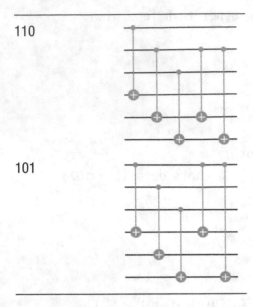

110

101

9.11

```
011 CX(0,0) CX(1,1) CX(2,2) CX(0,0) CX(0,1)
100 CX(0,0) CX(1,1) CX(2,2) CX(2,2)
101 CX(0,0) CX(1,1) CX(2,2) C(0,0) C(0,2)
110 CX(0,0) CX(1,1) CX(2,2) C(1,1) C(1,2)
```

9.12

```
# n-qubit version for Simon's oracle
def oracle (s):
  # reverse b for qiskit's qubit ordering
  s   = s[::-1]
  n   = len(s)
  qc  = QuantumCircuit(n * 2)

  # all 0s, so just exit
  if '1' not in s:
    return qc

  # index of first non-zero bit in s
  i = s.find('1')
```

```
for q in range(n):
  # Copy; |x>|0> -> |x>|x>
  qc.cx(q, q+n)

  # |x> -> |s.x> if q(i) == 1
  if s[q] == '1':
    qc.cx(i, (q)+n)
return qc
```

9.13

X, Y, H. Phase gates such as S-gate, T-gate and CPHASE-gate are not Involutory.

9.14

Pi/4

9.15 Applying Hadamard gate to all qubits in the given quantum circuit will put qubits in an equiprobable state (equal probability of being in all states).

b) qc.h(0) qc.h(1)

9.16

a) QuantumCircuit(3,3)

9.17

Option 1 is swap gate.

Option 2 is ccnot gate or toffoli gate.

Option 3 is controlled Z-gate.

Option 4 is controlled T-gate.

9.18

a,b

9.19

c. A = Identity, b = Y gate.

9.20

a) ([0, -i],[i,0]) This is the Y gate.

Index

© Vladimir Silva 2024
V. Silva, *Quantum Computing by Practice*, https://doi.org/10.1007/978-1-4842-9991-3

Printed in the United States
by Baker & Taylor Publisher Services